Impact of Variability in Carbamazepine Raw Materials on Drug Release

Inauguraldissertation

zur
Erlangung der Würde eines Doktors der Philosophie
Philosophisch-Naturwissenschaftlichen Fakultät
der Universität Basel

vorgelegt von

Felicia Flicker

Unterschächen (Uri)

Leiter der Arbeit: PD Dr. Gabriele Betz
Fakultätsvetretung: Prof. Dr. Matthias Hamburger
Departement Pharmazeutische Wissenschaften

Von der Philosophisch-Naturwissenschaftlichen Fakultät angenommen.

Der Dekan:
Basel, 27. Juni 2011   Prof. Dr. Martin Spiess

This dissertation is available as a free download from http://edoc.unibas.ch/diss/DissB_9729

Copyright © 2012 Felicia Flicker

This work is licensed under a Creative Commons Attribution-NonCommercial-NoDerivs 3.0 Unported License.

You are free:

to Share — to copy, distribute and transmit the work.

Under the following conditions:

Attribution — You must attribute the work in the manner specified by the author or licensor
 (but not in any way that suggests that they endorse you or your use of the work).

Noncommercial — You may not use this work for commercial purposes.

No Derivative Works — You may not alter, transform, or build upon this work.

Under the understanding that:

Waiver — Any of the above conditions can be waived if you get permission from the copyright holder.

Public Domain — Where the work or any of its elements is in the public domain under applicable law, that status is in no way affected by the license.

Other Rights — In no way are any of the following rights affected by the license:
 Your fair dealing or fair use rights, or other applicable copyright exceptions and limitations;
 The author's moral rights;
 Rights other persons may have either in the work itself or in how the work is used, such as publicity or privacy rights.

Notice — For any reuse or distribution, you must make clear to others the license terms of this work. The best way to do this is with a link to this web page.

 This is a human-readable summary of the Legal Code (the full license).
http://creativecommons.org/licenses/by-nc-nd/2.0/legalcode

ISBN 978-1-105-68249-0
First Edition, May 2012

Acknowledgments

I wish to express my gratitude to PD Dr. Gabriele Betz for giving me the opportunity to prepare my thesis at the Industrial Pharmacy Lab, Department of Pharmaceutical Sciences, University of Basel. Gabi, thank you for your excellent support, your positive energy, and encouragement.

My gratitude goes to Prof. Dr. Matthias Hamburger who kindly accepted the faculty responsibility for my thesis.

I am grateful to Prof. Dr. Anna Seelig who kindly accepted to be the second reviewer of this thesis and I thank Prof. Dr. de Capitani for accepting to chair the examination.

My thanks go to the Senglet Stiftung for financing my work.

I wish to thank my master student Veronika A. Eberle for her valuable work and for sharing a wonderful time.

I would like to thank Evi Bieler of Microscopy Center at Biozentrum for providing me with SEM images of my samples.

I would also like to thank Prof. Willem B. Stern and Prof. Christian de Capitani of the Department of Geosciences for allowing me to use the powder X-ray diffractometer and for their kind help.

My thanks go to the mechanical workshop for providing me with the made to measure tools for my modified dissolution system. My thanks also go to our laboratory assistant Stefan Winzap for his helpful presence.

My gratitude goes to Dr. Selma Šehić. I was honored to build my thesis based on her valuable work and I am grateful for her continuous support. I further thank for the thermometric analyses performed at her company Bosnalijek.

I owe my deepest gratitude to my dear colleagues Dr. Krisanin Chansanroj, Dr. Imjak Jeon, Dr. Muhanned Saeed, Dr. Ivana Vejnović, Dr. Murad Rumman, Dr. Elaine Darronqui, Dr. Sameh Abdel-Hamid, Dr. Firas Alshihabi, Dr. Nicolaos Gentis, Branko Vranić, Lizbeth Martínez, Abdoulaye Theophile Sebgo, Ylber Qusaj, Miki Yamashita, Yuya Yonezawa, Abhishek Kumar, and Sakine Tuncay for many fruitful discussions, exchange of laboratory and writing skills, and above all, for the many cheerful laughters and for sharing the unique "common spirit" of IPL.

I am deeply grateful to my parents and my sisters Anna, Rosa, and Odilia who have unconditionally supported me over all those years.

Above all, I thank my love, Adrian Lienhard, for his continuous support and for sharing many unforgettable moments.

Felicia Flicker
May 14, 2012

To Adrian

Summary

Variability in raw materials presents a challenge for pharmaceutical companies. The varying physicochemical properties can critically influence drug release and bioavailability of the final dosage form. Therefore, a strategy to control this variability is required. In this study the well-established antiepileptic drug carbamazepine (CBZ) was selected as the model drug as it presents one example where variability in raw materials has been linked to bioinequivalence and clinical failures. CBZ shows poor solubility, low potency, and a narrow therapeutic index. Furthermore, CBZ exhibits at least four polymorphic forms and it transforms into the less soluble CBZ dihydrate in water.

The purpose of this work was to study the impact of variability in CBZ samples of four different suppliers on the drug release and to suggest a strategy to deal with the sample variability. Thus, the CBZ samples were characterized at preformulation as well as at a formulation level.

Polymorphism and morphology of CBZ samples were analyzed by differential scanning calorimetry, X-ray powder diffraction, sieve analysis, and scanning electron microscopy. CBZ samples were characterized by a unidirectional dissolution method measuring disc intrinsic dissolution rate (DIDR) of CBZ raw material and initial drug release in presence of the tablet fillers microcrystalline cellulose (MCC) and mannitol (30–90% drug load). Furthermore, CBZ samples were recrystallized in 1% polyvinylpyrrolidone ethanol solutions as an approach to reduce the sample variability. At the formulation level, a high-dose CBZ tablet was developed with the aim of a tablet formulation that is robust towards the variability in CBZ samples and that conforms to the USP requirements of CBZ tablets for immediate release. Therefore, the superdisintegrant crospovidone (CrosPVP) and the dry binder hydroxypropyl cellulose (HPC) were used, as both are reported to inhibit transformation to CBZ dihydrate. The tablet filler was MCC.

All CBZ samples were of p-monoclinic form but differed in their polymorphic purity, particle size, morphology, and intrinsic dissolution rate. The DIDR profiles showed high variability among the CBZ samples. Two inflection points could characterize individual transformation behavior of anhydrous CBZ to CBZ dihydrate. Presence of MCC reduced drug release variability. Recrystallizing CBZ resulted in strongly reduced variability in dissolution and tablet strength and the transformation to

CBZ dihydrate was inhibited. However, particle size and morphology could not be controlled and drug release from binary mixtures with MCC presented deviation for one of the recrystallized CBZ samples. For the tablet formulation the optimal condition was with 6% HPC and 5% CrosPVP, where tablet properties of all CBZ samples were at least 70 N tablet hardness, less than 1 min disintegration, and within the USP requirements for drug release. Nonetheless, dissolution curves of the various CBZ samples differed. Excluding the additive sodium laurylsulfate required by the USP monograph and analyzing the optimized tablet formulation in water only, the dissolution curves of the various CBZ samples could not be distinguished anymore (ANOVA, $p > 0.05$).

The impact of variability in CBZ raw materials on the drug release could be characterized by an individual transformation behavior to the CBZ dihydrate. The applied unidirectional dissolution method can be suggested as a straightforward monitoring tool in preformulation studies conforming to the basic tenet of quality by design of FDA's PAT initiative. To allow a certain variability in CBZ raw materials, it is suggested to incorporate the excipients CrosPVP, HPC, and MCC into the design of a CBZ tablet formulation. The strategy proposed of how to control the variability in CBZ samples includes the monitoring at preformulation level combined with the design of a robust tablet formulation.

Contents

Acknowledgments iii

Summary vii

Abbreviations xiii

1 **Theoretical Introduction** 1
   - 1.1 Preformulation Studies 1
     - 1.1.1 Biopharmaceutics Classification System 2
     - 1.1.2 Preformulation Studies in the Context of the PAT-Initiative. 4
   - 1.2 Polymorphism 4
     - 1.2.1 Crystal Lattice 5
     - 1.2.2 Cocrystals and Solvates 6
     - 1.2.3 Transformation 7
     - 1.2.4 Polymorphism and Physical Properties 14
     - 1.2.5 Analytical Techniques 15
   - 1.3 Morphology 22
     - 1.3.1 Crystal Habit 22
     - 1.3.2 Crystallization 22
     - 1.3.3 Analytical Techniques 24
   - 1.4 Solubility and Dissolution Rate 26
     - 1.4.1 Disk Intrinsic Dissolution Rate 27
   - 1.5 Carbamazepine (CBZ) 28
     - 1.5.1 Transformation of Anhydrous CBZ to its Dihydrate Form . . 30
     - 1.5.2 Mechanical Properties of CBZ 32

		1.5.3	CBZ and Analytical Methods	32
		1.5.4	Techniques to Enhance CBZ Dissolution Rate	35
		1.5.5	Irregular Dissolution, Bioinequivalence, and Clinical Failures in CBZ Tablets	36
		1.5.6	CBZ Tablets Registered in Switzerland	37
	1.6	Formulation Studies		37
	1.7	Direct Compaction		37
	1.8	Choice of Excipients for Direct Compaction		38
		1.8.1	Filler-binder	39
		1.8.2	Disintegrant	41
		1.8.3	Lubricant	43
	1.9	Analysis of Tablet Properties		43
		1.9.1	Tablet Hardness	43
		1.9.2	Tablet Friability	44
		1.9.3	Tablet Porosity	44
		1.9.4	Disintegration Testing	44
		1.9.5	In Vitro Dissolution Testing	44
	Bibliography			47

Bibliography   49

2 Objectives   61

3 Original Publications   63

    3.1 Variability in Commercial Carbamazepine Samples – Impact on Drug Release 63

    Bibliography 78

Bibliography   83

    3.2 Effect of Crospovidone and Hydroxypropyl Cellulose on Carbamazepine in High-Dose Tablet Formulation 87

    Bibliography 103

Bibliography   105

4 Recrystallization Project   109

Bibliography 119

Bibliography   **121**

5 Dissolution Project with the Optimized CBZ Tablet Formulation   **123**
     Bibliography 125

Bibliography   **129**

6 Conclusions   **131**

Appendix   **133**
     6.1 Additional Information on Publication 1 and on the Recrystallization Project 133
         6.1.1 Further Analyses on CBZ Samples 133
         6.1.2 Development of the Unidirectional Dissolution Method . . 136
         6.1.3 Precision and Effect of Particle Size 140
         6.1.4 Repeatability 141
         6.1.5 Inflection Point in DIDR Profiles of CBZ Dihydrate? . . . 143
         6.1.6 Transformation of CBZ to Dihydrate – Contradictory Results in Literature 143
         6.1.7 Compact Hardness of Untreated and Recrystallized CBZ Samples 144
         6.1.8 Effects of Mannitol and MCC on CBZ 144
         6.1.9 UV Calibration 149
     6.2 Additional Information on Publication 2 and on the Dissolution Project 152
         6.2.1 SEM Images of the Excipients in the Tablet Formulation . 152
         6.2.2 UV Calibration 152
     Bibliography 153

Bibliography   **155**

Abbreviations

ANOVA	Analysis of Variance
API	Active Pharmaceutical Ingredient
a_W	Water Activity
BCS	Biopharmaceutics Classification System
CBZ	Carbamazepine
CBZ re	Recrystallized Carbamazepine
CrosPVP	Crospovidone = Cross-linked Polyvinylpyrrolidone
DSC	Differential Scanning Calorimetry
DIDR	Disk Intrinsic Dissolution Rate
FDA	Food and Drug Administration
HPC	Hydroxypropyl Cellulose
HPMC	Hydroxypropyl Methylcellulose
IDR	Intrinsic Dissolution Rate
MANOVA	Multivariate Analysis of Variance
MCC	Microcrystalline Cellulose
PAT	Process Analytical Technology
PEG	Polyethylene Glycol
PVP	Povidone = Polyvinylpyrrolidone
RH	Relative Humidity
rpm	rounds per minute
RSD	Relative Standard Deviation

SEM	Scanning Electron Microscopy
SLS	Sodium Laurylsulfate
STC	Sodim Taurocholate
SD	Standard Deviation
USP	United States Pharmacopoeia
XRPD	X-ray Powder Diffractometry

Chapter 1

Theoretical Introduction

1.1 Preformulation Studies

Preformulation studies assess the physicochemical and biopharmaceutical properties of a drug candidate. These properties show whether the drug candidate can be formulated and they hint at potential problems in the drug performance and stability (Wells and Aulton, 2007). A thorough understanding of the drug properties is not only effective to reduce drug development time and cost but is also crucial to the quality and safety of the drug formulation. Therefore, preformulation is part of the official requirements for investigational new drugs and new drug applications by the Food and Drug Administration (FDA). Also commercial requirements on drug delivery and dosage form can be a driving factor for preformulation studies (Carstensen, 2002).

An overview on the studies performed in preformulation is shown in Table 1.1. For the very first physicochemical studies the synthesis of a drug is not at its final scheme. After scale-up less impurities and more relevant data can be obtained, only then preformulation studies with higher precision make sense. Also studies on powder flow, compaction properties, and excipient compatibility are important characteristics of a drug, however, they are often analyzed at a later stage as only small amounts (mg) of the new drug are available at earlier stages (Carstensen, 2002).

In case of abbreviated new drug applications (ANDAs), also called generics, the preformulation studies are less intensive (FDA, 2007). Nonetheless, physicochemical properties of a drug can vary with the source and this variation may lead to irregular dissolution behavior and clinical failures of the drug formulation (Wang et al., 1993; Meyer et al., 1992, 1998; Davidson, 1995; Jung et al., 1997; Lake et al., 1999; Mittapalli et al., 2008).

Table 1.1: Preformulation studies on a new chemical entity (NCE) according to Wells and Aulton (2007).

Method	Characterization
Spectroscopy	Simple UV assay
Solubility	Phase solubility, purity
— aqueous	Intrinsic solubility, pH effects
— pK_a	Solubility control, salt formation
— salts	Solubility, hygroscopicity, stability
— solvents	Vehicles, extraction
— partition coefficient	Lipophilicity, structure activity
— dissolution	Biopharmaceutics
Melting point	DSC-polymorphism, hydrates, solvates
Assay development	UV, HPLC
Stability	
— in solution and solid state	Thermal, hydrolysis, oxidation, photolysis, metal iones, pH
Microscopy	Morphology particle size
Powder flow	Tablet and capsule formulation
— bulk density	
— angle of repose	
Compaction properties	Excipient choice
Excipient compatibility	

1.1.1 Biopharmaceutics Classification System (BCS)

Important guidance for the formulation scientist is the Biopharmaceutics Classification System (BCS) proposed by Amidon et al. (1995). It allows classifying drugs according to their solubility and permeability (Figure 1.1). Furthermore, also the dose has to be considered and therefore the FDA's Center of Drug Evaluation and

Research (FDA CDER, 2009) defined the limits of the cut-off values for the BCS classes including the drug dose. High solubility is defined as highest dose strength soluble in < 250 ml water over a pH range of 1.0–7.5 and high permeability is defined as the absorption of > 90% of an administered dose. The permeability is determined by a transport model and human permeability results. The four BCS classes and their implications on the drug formulation are discussed below (He, 2008):

- Class I: High solubility and high permeability of the drug, formulation for immediate drug release can be achieved without major challenge.

- Class II: An increasing number of the new chemical entities (NCEs) belong to this class of low solubility and high permeability, where the bioavailability is dissolution rate controlled. The challenge here is to overcome the low solubility.

The following options are available to improve solubility and thereby shift the drug from class II to I: (1) salt formation, (2) particle size reduction, e.g., nano particles, (3) metastable forms or amorphous state, (4) solid dispersion, (5) complexation, e.g., with cyclodextrins, (6) lipid based formulations, e.g., self-emulsifying drug delivery systems, and (7) inhibition of precipitation / crystallization in the gastrointestinal tract (He, 2008).

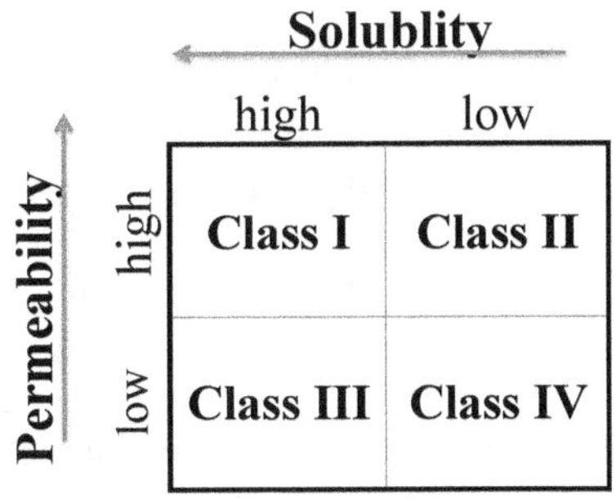

	Solublity	
Permeability	high	low
high	Class I	Class II
low	Class III	Class IV

Figure 1.1: Biopharmaceutics Classification System (BCS).

- Class III: High solubility but low permeability of the drug, the oral route of administration can only be an option if prodrugs or permeability enhancers are available.

- Class IV: Solubility as well as permeability problems have to be addressed, it is most difficult to achieve formulation from class IV drugs. Alternative delivery routes, such as intravenous administration are often the only solution. In general, poor permeability is rarely overcome by formulation approach.

The BCS has been refined to classify drugs according to their stage of development (Papadopoulou et al., 2008). Whereas new chemical entities can be classified based on the solubility-dose ratio and permeability estimates, marketed drugs can be classified by mean dissolution time (MDT) and mean permeation time (MPT) of the gastrointestinal wall. Zakeri-Milani et al. (2009) were able to classify drugs according to the BCS by intrinsic dissolution rate, which is much faster obtained compared to the drug solubility.

Besides the BCS, also the metabolic or chemical instability of a drug may ask for special formulation measures (He, 2008). Acid sensitive drugs could be protected by an enteric coating, whereas gastrointestinal metabolism can be minimized by coadministration with enzyme inhibitor. Also the effect of the P-gp efflux pump can be minimized by coadministration of an inhibitor. A further approach is to use carrier-mediated transport by designing a prodrug that is a substrate for the transporter.

1.1.2 Preformulation Studies in the Context of the PAT-Initiative.

Preformulation studies have to be placed in the context of FDA's initiative on Process Analytical Technology (PAT). It presents "A Framework for Innovative Pharmaceutical Development, Manufacturing, and Quality Assurance" (FDA, 2004). The PAT initiative aims for scientific understanding of the pharmaceutics of a drug candidate. Beginning with the therapeutic objective, the patients, the route of administration, the pharmacokinetics and pharmacodynamics, followed by the physicochemical and biopharmaceutical properties, the drug formulation, and finally by the manufacturing processes. A drug formulation and process should be achieved by design and its critical parameters should be known in order to obtain a robust formulation with risk-based processing and built-in quality. The PAT initiative has the final aim of real time release of a product with predefined quality. Therefore, a wide range of tools are necessary, tools for multivariate designs, to analyze and control the processes, and tools for the data and knowledge management to ensure continuos improvement.

In the following sections the preformulation topics polymorphism, morphology, solubility and dissolution rate are discussed in detail. In Section 1.5, the model drug carbamazepine (CBZ) is presented from several preformulation and formulation aspects. The choice of excipients is addressed in the Section 1.8 on formulation studies.

1.2 Polymorphism

The term polymorphism has its origin in the greek words poly ($\pi o \lambda \acute{v}$) for "many/much" and morphē ($\mu o \rho \varphi \acute{\eta}$) for "form" (Hilfiker, 2006). In chemistry, polymorphism is the ability of a single molecule to crystallize in more than one distinct crystal architecture (Rustichelli et al., 2000). This is analog to allotropy, the chemical term for elements that crystallize in more than one crystal form. Among polymorphic forms there is always one crystal form that is the thermodynamically most stable form. The comparatively less stable form is called *metastable* and the absence of a crystal structure, meaning no long-range order, is the *amorphous* state (Yu, 2001). An estimate of 90% organic compounds can exhibit multiple solid states including crystalline polymorphs (about 50%), solvates and noncrystalline forms (Stahly, 2007).

1.2.1 Crystal Lattice

The crystal lattice is the geometric arrangement of molecules in a crystal architecture. The lattice parameters are described by the axes a, b, and c and the angles α, β, and γ of a unit cell, the smallest unit of a crystal. There are seven different crystal systems, where the cubic form presents the simplest cell with all axes of same length and all angles at 90°.

Some unit cells allow a further molecule to be positioned at the center, the base or at the face of the unit cell and a total of 14 different lattices can be formed (Rodrìguez-Hornedo et al., 2007). Table 1.2 presents an overview on the possible crystal lattices and Figure 1.2 shows the p-monoclinic cell (monoclinic, simple) of carbamazepine as an example.

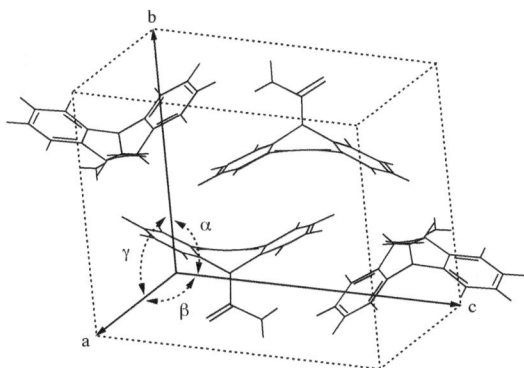

Figure 1.2: Unit cell of p-monoclinic carbamazepine (Rodrìguez-Hornedo et al., 2007).

Crystal lattices can be further distinguished by the way they are formed. *Packing polymorphism* refers to rigid molecules of fixed conformation that are packed in different arrangements, while *conformational polymorphism* refers to flexible molecules packed with different conformations (Aaltonen et al., 2009).

Table 1.2: The seven crystal systems and their lattice parameters (Patterson and Bailey, 2007; Cullity and Stock, 2001).

System	lattice parameters	Bravais lattices
Triclinic	$\alpha \neq \beta, \beta \neq \gamma, \alpha \neq \gamma$	simple (p)
	$a \neq b, b \neq c, a \neq c$	
Monoclinic	$\alpha = \gamma = 90°, \beta \neq \alpha$	simple (p)
	$a \neq b, b \neq c, a \neq c$	base-centered (c)
Orthorhombic	$\alpha = \beta = \gamma = 90°$	simple (p)
	$a \neq b, b \neq c, a \neq c$	base-centered (c)
		body-centered (i)
		face-centered (f)
Tetragonal	$\alpha = \beta = \gamma = 90°$	simple (p)
	$a = b \neq c$	body-centered (i)
Trigonal	$\alpha = \beta = \gamma \neq 90°, \leq 120°$	simple (p)
	$a = b = c$	
Hexagonal	$\alpha = \beta = 90°, \gamma = 120°$	simple (p)
	$a = b \leq c$	
Cubic	$\alpha = \beta = \gamma = 90°$	simple (p)
	$a = b = c$	body-centered (i)
		face-centered (f)

1.2.2 Cocrystals and Solvates

Cocrystals or solvates are formed if a drug crystallizes together with guest molecules in the unit cell. In cocrystals the different molecules are in solid state, whereas in solvates they are in liquid state at ambient conditions. Also cocrystals and solvates can crystallize into different polymorphic forms (Aaltonen et al., 2009).

Pharmaceutical Hydrates

Solvates with water are called hydrates. Crystallization of a drug molecule including molecules of water leads to a higher stability in aqueous conditions. However, some hydrates are also kinetically stable at ambient conditions, which makes it possible to choose the hydrate form for the drug development (Aaltonen et al., 2009). The water molecule is of small size and it is capable of multidirectional hydrogen bonds. Hydrates can be classified according to the arrangement of the water molecules (Vippagunta et al., 2001). Class I hydrates show isolated water molecules. Class

II hydrates are channel hydrates, with either a clear two-dimensional arrangement of water molecules (class IIa) or with nonstociometric amount of water, where the lattice expands at higher relative humidity to include more water molecules. The class III hydrates show ion-associated water molecules, this class of hydrates can only form drug molecules with metal-ions.

1.2.3 Transformation

As soon as a drug exhibits multiple solid-states the question of transformation arises. Thermodynamics determine the relative stability of a solid-state and show which transformation can take place at what conditions. The duration of a transformation, however, is determined by the kinetics. Furthermore, a solid-phase does not transform without molecular recognition. Therefore, transformation is governed by the

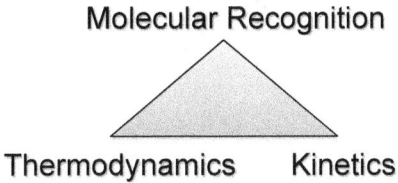

Figure 1.3: Schematic diagram of the influences governing the solid-phase transformations (Rodrìguez-Sponga et al., 2004).

three principles of thermodynamics, kinetics, and molecular recognition as shown in Figure 1.3 (Rodrìguez-Sponga et al., 2004).

Thermodynamics

The Gibbs's phase rule (Equation 1.1) describes the relationship between the different solid phases:

$$P + F = C + 2 \qquad (1.1)$$

where C is the degree of freedom and C and P are the number of components and phases that exist in equilibrium, respectively. In a system of one single substance ($C = 1$) and one polymorphic form ($P = 1$) the degree of freedom is $F = 2$. Therefore temperature and pressure can both vary without changing the number of polymorphic forms. However, two polymorphic forms of the same substance ($P = 2$, $C = 1$) can only coexist in equilibrium if either temperature or pressure is constant ($F = 1$). In this case a fixed transition temperature (T_t) exists at atmospheric pressure (Giron, 1995; Grant, 1999).

Polymorphic transformation is the transition of one polymorphic form (I) to another (II) at T_t. Depending on the melting point (T_m) of the forms two transition cases exist (Figure 1.4). If $T_t < T_m$, the transition is reversible and the polymorphic forms are enantiotropic. In this case form I is thermodynamically more stable below T_t, whereas above T_t form II is more stable. On heating, the transition is endothermic. In the other case, if $T_t > T_m$, form I melts before the transition to form II, the transformation of form II to I is exothermic and irreversible. The polymorphic forms are monotropic (Giron, 1995; Brittain, 1999). Enantiotropic and monotropic

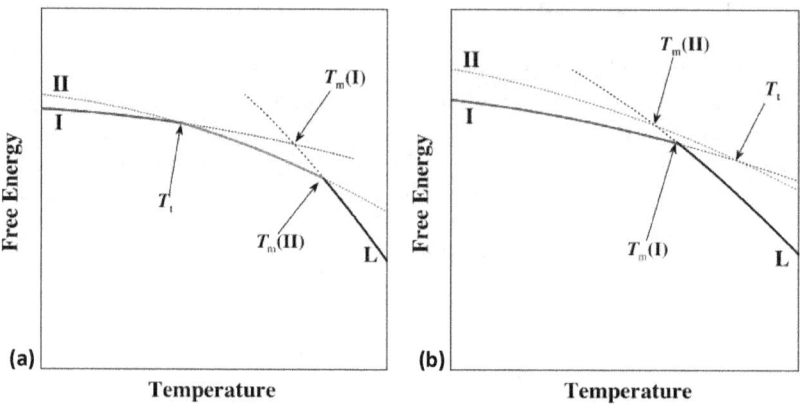

Figure 1.4: Free energy phase diagrams of polymorphic forms (I, II) in an enantiotropic (a) and monotropic (b) system. T_t and T_m are the temperature of transition and melting, and L is the melted phase (Zhang et al., 2004).

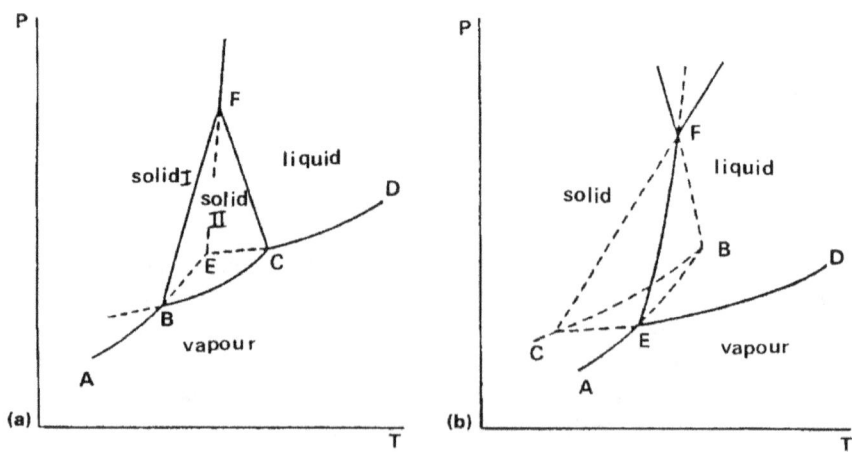

Figure 1.5: Temperature (T) – pressure (P) diagrams of a single substance with enantiotropic (a) and monotropic (b) transition behavior (Giron, 1995).

phase transition is not limited to the solid-solid phase transition but include also the solid-liquid and solid-gas transition. Phase diagrams of pressure versus temperature show the different transitions by equilibrium curves. Components with enantiotropic behavior (Figure 1.5a) present a liquid-vapor equilibrium curve (CD) that meets the two solid-vapor curves *after* the point of intersection with the solid-solid equilibrium curve. There is a solid I – solid II equilibrium curve (BF) and a reversible transition point I to II at a specific pressure. At the transition point, the free energy of the two forms is the same. AB shows the equilibrium of solid I – vapor, BC the equilibrium solid II –vapor, and FC the equilibrium of solid II – liquid phase. In case of components with monotropic behavior (Figure 1.5b) the liquid-vapor equilibrium curve (ED) crosses the solid I – vapor curve (AE) *before* the points B and C. AB shows the solid I – liquid equilibrium curve. The dashed curves represent equilibrium curves of the metastable polymorphic form (Giron, 1995).

Kinetics

The kinetics influencing transformation depend on the energy barriers in the system. The ability to undergo transformation and the relative thermodynamic stability of two polymorphic forms at constant pressure are given by the Gibbs free energy G [J] (Equation 1.2),

$$\Delta G = \Delta H - T\Delta S \qquad (1.2)$$

where H is the enthalpy [J], T the temperature [K], and S the entropy of system [JK^{-1}]. A difference in entropy ΔS reflects disorder and lattice vibration between two polymorphic forms and difference in enthalpy ΔH is related to structural or lattice energy. The difference in Gibbs free energy ΔG shows the relative stability of the system by three distinct situations: (1) ΔG is negative and transformation can take place spontaneously and transformation can continue as long as the free energy of the system is negative; (2) $\Delta G = 0$, the system is at equilibrium, both phases have the same free energy and no transformation occurs; (3) ΔG is positive and transformation is not possible under the fixed conditions (Rodrìguez-Sponga et al., 2004).

The schematic diagram in Figure 1.6 shows the balance between kinetic and thermodynamic factors by the hypothetical transition of two different solid phases I and II, where phase I is the the more stable and less soluble form ($G_I < G_{II}$). The initial state (G_X) presents a negative ΔG for both forms, giving the thermodynamic driving force of the transition. Although the absolute ΔG value is higher for solid phase I, the kinetic barrier is smaller for form II ($G_I^* > G_{II}^*$) and transition my be in favor of form II (Rodrìguez-Sponga et al., 2004).

Transformation and the Underlying Mechanisms

Other than the above mentioned mechanisms of solid-solid (solid-state) and the solid-liquid (melt) transition, also solution and solution-mediated transition are mechanisms leading to transformation. In the latter mechanisms the influence of solvents and water are the critical parameters. The four types of phase transition are discussed below (Zhang et al., 2004):

- *Solid-state transition* occurs by (1) molecular loosening, followed by (2) an intermediate solid solution, where (3) the new solid-phase can nucleate and (4) grow (Vippagunta et al., 2001). The process can be induced by a change in temperature, pressure, and relative humidity or by crystal defects and impurities.

- *Melt transition* occurs through the cooling of melted solid, another form can crystallize. The final solid phase is determined by the rates of cooling, nucleation, and crystal growth. Furthermore, impurities and excipients influence the crystallization process.

- *Transition in solution* only occurs in the fraction of the drug that is dissolved. After removing the solvent the substance may be present in a different solid state often resulting in polymorphic mixtures.

- *Solution-mediated transition* occurs in direction of the phase of higher stability. The mechanism is by (1) dissolution of the metastable phase, (2) nucleation of the stable phase, and (3) growth of the stable phase. This transformation is usually faster than the solid-state transformation as the

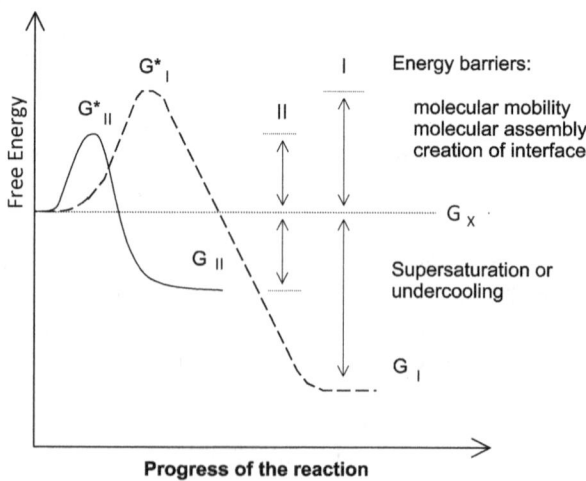

Figure 1.6: Schematic diagram with the hypothetical transition reactions of two different solid phases I and II; G_X is the initial free energy, G^* and G are the maximal and minimal energy states of each solid (Rodrìguez-Sponga et al., 2004).

mobility necessary for the rearrangement is provided by the solution. In general, nucleation or crystal growth are the rate-controlling steps of the transformation. The kinetics of both are affected by solubility difference between the phases, contact surface (particle size), as well as temperature and agitation of the solution, and soluble impurities and excipients. Nucleation is further influenced by the solubility and crystal growth by the solid/solvent ratio (Vippagunta et al., 2001; Zhang et al., 2004).

Depending on the resulting solid phase the types of transition can be further classified into polymorphic transition, hydration/dehydration, and vitrification. Table 1.3 provides an overview (Zhang et al., 2004).

- *Polymorphic* transition is explained in the paragraph on polymorphic transformation above. It is further interesting to note, that polymorphic transformation often occurs over a sequence of phase transitions. A typical transition sequence in manufacturing could start with hydration and be followed by dehydrate or begin with vitrification an proceed with a crystallization step.

- *Hydration and dehydration* are the transitions between crystalline anhydrates and hydrates, and between lower and higher hydrates. The transitions depend on temperature, pressure, and on the water activity. Water activity and transformation is discussed below.

Table 1.3: Classification of phase transitions by type of transition and resulting solid phase (Zhang et al., 2004).

Solid-state	Polymorphic transition
	Hydration/dehydration
	Amorphous crystallization/ vitrification
Melt	Polymorphic transition
	Vitrification
Solution	Polymorphic transition
	Hydration/dehydration
	Amorphous crystallization/ vitrification
Solution-mediated	Polymorphic transition
	Hydration/dehydration
	Amorphous crystallization
	Only from the metastable phases to the stable phases

- *Vitrification and amorphous crystallization* is the transition between crystalline solid and amorphous phase. Amorphous phase can convert to crystalline solid over all types of transition. If temperature is above the glass transition and in presence of moisture (plasticizer and anti-solvent) kinetics drive the system towards a faster crystallization as both cases result in increased mobility. The transition of crystalline to amorphous phase (vitrification) is most likely to occur by a fast cooling or solvent evaporation, as the crystal nucleation and growth are the slower processes. Furthermore, also mechanical stresses on hydrates can easily lead to vitrification.

Water Activity

Water plays an important role in phase transition, especially if they are by solution or solution-mediated. Khankari and Grant (1995) have described the thermodynamics of the hydration - dehydration process. The equilibrium of the two phases is given by the Equations 1.3 and 1.4,

$$A_{(S)} + mH_2O \xrightleftharpoons{K_h} A \cdot mH_2O_{(S)} \qquad (1.3)$$

$$K_h = \frac{a[A \cdot mH_2O_{(S)}]}{a[A_{(S)}]a[H_2O]^m} \qquad (1.4)$$

where m is the number of moles of water taken up by 1 mol of anhydrate, $a[A_{(S)}]$, $a[A \cdot mH_2O_{(S)}]$, and $a[H_2O]$ are the thermodynamic activity of the anhydrate, the hydrate, and the water, respectively. In which direction the transformation occurs is indicated by the equilibrium constant K_h (Khankari and Grant, 1995).

$$K_h = a[H_2O]^{-m} \qquad (1.5)$$

The equilibrium constant can be simplified as to Equation 1.5, if hydrate and anhydrate are pure solids and are considered as one unit expressing thermodynamic activity (Rodrìguez-Sponga et al., 2004).

Water activity is a key parameter in the transformation process, the driving force for the transition increases with the water activity a_w (Equation 1.6), where x_w is the mole fraction of water in the system and γ_w the activity coefficient (Rodrìguez-Sponga et al., 2004).

$$a_w = \gamma_w x_w \qquad (1.6)$$

Water activity can be monitored by measuring the equilibrium relative humidity (ERH) of a system (Equation 1.7),

$$ERH = \frac{p}{p_0} = a_w \cdot 100 \qquad (1.7)$$

where p is the partial vapor pressure and p_0 the vapor pressure of pure water. All parameters are temperature dependent (Wrolstad et al., 2005).

Stability Issues in Pharmaceutical Solids

The stability of a system correlates well with the water activity, whereas the amount of water may not always relate to the stability (Rockland and Beuchat, 1986). The molecular bases of moisture effects on drug stability is discussed in a review article by Ahlneck and Zografi (1990). In case of excipient-drug interactions there are two mechanisms describing how water can be involved. The first mechanism is a redistribution of water by a vapor phase from the excipients to the drug and the second mechanism is by direct physical contact of sorbed water located between the drug and the excipients. The amount of freely available water in a system depends on temperature and pressure and is expressed by the water activity. If water adsorbs to the drug its viscosity can change, molecular mobility increases, and as a consequence leads to higher reactivity and instability. An important study on stability issues related to moisture has been published by Dürig and Fassihi (1993). They studied the destabilizing effects of moisture and elevated temperature on the excipient-drug interaction. Interaction of the moisture-sensitive drug pyridoxal hydrochloride was tested with the following excipients: colloidal silicone dioxide, anhydrous lactose, modified lactose (Ludipress®, BASF), corn starch, microcrystalline cellulose (Avicel® PH101), methylcellulose, ethylcellulose, polymethacrylate derivates (Eudragit® RSPM), stearic acid, and magnesium stearate. They suggested that hygroscopic excipients such as microcrystalline cellulose (MCC) can function as "moisture scavengers". Although MCC adsorbs a high amount of water, it appears to bind strongly to the hydroxy groups in the cellulose, water was not freely available for the interaction, thus the system presented a low water activity. Colloidal silicone dioxide and the various cellulose derivates showed strong destabilizing effect.

A critical point are crystal defects in form of local disorder; they present an activated state. These regions are able to take up more water compared to the intact crystal surface and can thus show higher molecular mobility and less stability. Therefore, special care has to be taken in many pharmaceutical processes causing crystal defects (Ahlneck and Zografi, 1990). The following pharmaceutical processes have been associated with possible phase transitions (Zhang et al., 2004):

- Size reduction by milling
- Size enlargement by granulation, i.e., wet granulation followed by drying, dry granulation, melt granulation, spray-dry and freeze-dry methods
- Granulation milling/sizing and blending; minimal risk
- Compaction and encapsulation; solid-solid transitions with minimal risk
- Coating by coating pans or fluid bed

1.2.4 Polymorphism and Physical Properties

The various solid phases exhibit different physical properties (Aaltonen et al., 2009), an overview is given in Table 1.4. The question arrises, what is the impact of polymorphism on the physical properties of the drug and can they be predicted?

Table 1.4: Physical properties that vary among the different solid phases of a drug, table taken from Aaltonen et al. (2009).

Packing properties	Unit cell volume (crystalline forms only), Density, Hygroscopicity, Refractive index
Thermodynamic properties	Enthalpy, Entropy, Free energy, Melting point, Solubility
Spectroscopic properties	Electronic transitions (UV-Vis spectra), Vibrational transitions (IR and Raman spectra), Rotational transitions (far-Infrared spectra), Nuclear spin transitions (NMR spectra)
Kinetic properties	Dissolution rate, Rates of solid-state reactions, Stability
Surface properties	Surface free energy, Interfacial tensions, Crystal habit
Mechanical properties	Hardness, Tensile strength, Compactibility, Tableting, Flowability

Summers et al. (1977) proposed a semi-empirical rule to predict the effect of solid-state on its compressibility and bonding strength by the crystal packing of the polymorphic form. The more stable form is expected to have higher packing density and stronger inter-particle bonds but with less ability to deform. Thus the weaker tablets are obtained. The density rule by Burger and Ramberger (1979) also states that the polymorphic form with the less densely packed molecules is the less stable form. However, the density rule only applies for polymorphs where the molecular packing is dominated by van der Waals interaction (Rodrìguez-Sponga et al., 2004).

Moreover, mechanical properties may only be predicted taking also the crystal morphology into account (Roberts et al., 2000).

The effect of polymorphism on the final dosage form depends on the absolute difference among the physical properties of a polymorphic drug, the drug load, and the influence of the manufacturing on polymorphic stability. To ensure a reproducible bioavailability of the dosage form, the thermodynamically most stable polymorph should be selected. The most stable polymorph shows the lowest risk of transformation during manufacturing and storage under real-world conditions. Only if a drug cannot be crystallized or if tremendous advantages in dissolution are observed also the amorphous or the metastable state may be selected. However, metastable and amorphous solids need to be kinetically stabilized by excipients or special processing (Singhal and Curatolo, 2004). Also the hydrate may be selected for the dosage formulation, however also here mechanical, thermal, and chemical stresses can induce phase transition and thereby threaten the stability of the final drug formulation (Rodrìguez-Sponga et al., 2004).

1.2.5 Analytical Techniques

Several analytical techniques are available to study the solid-state properties of polymorphic drugs. They can be grouped according to the aspect of solid-state properties they characterize (Tables 1.5–1.8). Table 1.5 gives an overview on crystallographic analyses by X-ray diffraction and Table 1.6 shows the thermal methods that allow analyzing phase transitions. The spectroscopical methods analyzing molecular motion, chemical environment of the molecules, and intramolecular and lattice vibrations are presented in Table 1.7, and the microscopical methods to study morphology are shown in Table 1.8 together with other methods such as the density measurement. In general, a combination of techniques is selected, as there is no superior method for all solid-state properties (Chieng et al., 2011). Although the tools available for solid form screening have evolved drastically over the past decade, they do not replace basic thermodynamic understanding in successful data analysis (Aaltonen et al., 2009). The major challenges remain the differentiation between two structurally similar polymorphs, the quantitative analysis of a single-solid state in mixtures of polymorphs, and the quantitative determination of amorphous or crystalline state in a system (Vippagunta et al., 2001).

X-ray Powder Diffraction (Suryanarayanan and Rastogi, 2007)

X-ray powder diffraction (XRPD) is the "gold" standard for qualitative and quantitative solid-phase identification. The specific diffraction pattern of a pharmaceutical solid is generated by directing X-rays at a single crystal or at a flattened surface of a tightly packed powder sample, the rays are then diffracted from the crystal lattice and scattered in all directions. At some directions the scattered beams become in phase

Table 1.5: Analytical methods to study the crystalline structure; (+) advantages and (−) disadvantages (Chieng et al., 2011; Zhang et al., 2004).

Crystallography − X-ray Diffraction	
X-ray powder diffraction (XRPD)	(+) "gold" standard for phase identification
	(+) qualitative and quantitative
	(+) non-destructive
	(−) preferred orientation
Single crystal X-ray diffraction	(+) ultimate phase identification
	(+) solves the crystal structure
	(+) non-destructive
	(−) a single crystal of > 0.1 mm necessary
Small angle X-ray scattering	(+) probes structures in nm–μm range
	(+) non-destructive
	(−) relatively long data acquisition time
	(−) needs advanced interpretation of the data

and are reinforced. This phenomenon is described by the *Bragg law* (Equation 1.8),

$$n\lambda = 2d \sin\theta \qquad (1.8)$$

where λ is the angle of incident on the sample, n the order of reflection as integer number, and d the distance between the successive planes in the crystal lattice [Å]. Each crystalline compound shows a distinct peak pattern and also solvates can be identified, as long as their crystal lattices differ. Disordered lattices (amorphous powder) are represented in the form of a characteristic broad halo in the XRPD pattern.

A powder sample often exists as an intermediate state of ordered (crystalline) and disordered lattice. By Equation 1.9 the proportion of crystalline state can be determined as the *degree of crystallinity* (x_{cr}) in [%],

$$x_{cr} = \frac{I_c 100}{I_c + \frac{qI_a}{p}} \qquad (1.9)$$

where I_c and I_a are the intensities measured for the crystalline and amorphous state, and p and q are proportionality constants. The intensities are best measured as integrated line intensity (area under the curve) and not as peak intensity (peak height).

Drug − excipient interactions are easily detected by XRPD if all components are crystalline. The diffraction pattern of a powder mixture is the summation of

Table 1.6: Analytical methods to study phase transitions; (+) advantages and (−) disadvantages (Chieng et al., 2011; Zhang et al., 2004).

Thermal Methods	
Differential scanning calorimetry (DSC)	(+) information on phase transition
	(+) information on interaction with excipients
	(+) qualitative and quantitative
	(+) small sample size
	(−) no information on the nature of the transformation
	(−) no separation of thermal events at the same temperature
Modulated temperature DSC	(+) improve clarity of small (i.e., T_g) and overlapping thermal events
	(−) more experimental variables (i.e., amplitude and period settings)
	(−) relative long data acquisition time
Thermogravimetric analysis, Dynamic vapor sorption	(+) quantitative information on stoichiometry of solvates/hydrates
	(+) small samples size
	(−) interference of water-containing excipients
	(−) samples is destroyed during analysis
Isothermal microcalorimetry	(+) high sensitivity
	(+) qualitative and quantitative
	(+) non destructive
	(+) stability studies directly under the storage conditions
	(−) low specificity
	(−) large amount of samples necessary
Solution calorimetry	(+) qualitative and quantitative
	(−) low specificity
	(−) large amount of samples necessary
	(−) sample not recovered
	(−) long measurement time

Table 1.7: Analytical methods to study solid phase at particulate level; (+) advantages and (−) disadvantages (Chieng et al., 2011; Zhang et al., 2004).

Molecular Motion - Vibrational Spectroscopy	
Mid-IR:	(+) small sample size,
FT-IR (Fourier transformed infrared),	(+) relatively fast methods,
DRIFTS (diffused reflectance infrared transmission spectroscopy),	(+) spectral libraries available
	(−) possible solid-state transformation in sample preparation
ATR (attenuated total reflectance)	(−) environmental humidity interference
	(+) no sample preparation (ATR)
Raman	(+) small sample size, no preparation,
	(+) non-destructive
	(−) local heating of sample
	(−) photodegradation
Near infrared (NIR)	(+) non-invasive and fast
	(+) no sample preparation
	(−) low sensitivity, chemometrics
Chemical Environment at Molecular Level	
Solid-state nuclear magnetic resonance	(+) non-destructive phase identification
	(+) qualitative and quantitative,
	(+) no calibration
	(−) relative long data acquisition time
	(−) relatively expensive
Intramolecular and Lattice Vibrations	
Far-infrared:	(+) small samples (5–40 mg)
terahertz pulsed spectroscopy	(+) fast data acquisition (milliseconds)
	(−) spectrum affected by water
	(−) relatively expensive
	(−) pellet compression needed

Table 1.8: Analytical techniques to study morphology, water content, surface area, and true density; (+) advantages and (−) disadvantages (Chieng et al., 2011; Zhang et al., 2004).

Morphology	
Polarized light microscopy (PSM)	(+) small sample size
	(+) easy to use
	(+) very little sample preparation
	(−) no quantitative information
PSM with hot/cryo/freeze drying stage	(+) temperature variability
	(−) careful sample preparation needed
Scanning electron microscopy (SEM)	(+) high resolution, small sample size
	(−) sample preparation, vacuum setting
Bulk level/Other	
Karl Fischer titration	(+) water content (adsorbed or hydrate) with high sensitivity
	(+) rapid analysis
	(−) sample must dissolve in the medium
	(−) sample size > 50 mg is preferred
Brunauer, Emmett and Teller method	(+) non-destructive
	(+) simple and straightforward method
	(−) degassing step required
	(−) sample size 50–100 mg
Density (gas pycnometer)	(+) simple and straightforward method
	(+) non-destructive
	(−) degassing step
	(−) sample size > 50 mg is preferred

diffraction pattern of each individual component. Interaction is visible as extra peaks or amorphous product is detected by the broad halo (Suryanarayanan and Rastogi, 2007). XRPD can also give the percentages of each component although the limit of detection varies strongly and can be as high as 15% (Vippagunta et al., 2001).

Most pharmaceutical compounds are organic molecules and they can be a challenge to the XRPD measurement because organic compounds tend to crystallize with crystal lattice of large d-spacings (accuracy in XRPD ↓), with unit cells of lower symmetry (complex patterns), and with low mass attenuation coefficients (beam transparency errors ↑).

A further challenge is the sample preparation. Non-random distribution, referred to as preferred orientation, leads to inaccurate peak pattern because the system sees a secondary cell structure. Preferred orientation can be reduced by particle size reduction. However, grinding may result in reduced crystallinity and in small particles, both leading to a peak broadening in the XRPD pattern.

Differential Scanning Calorimetry (Clas et al., 2007).

Differential Scanning Calorimetry (DSC) is the analytical technique of choice to determine melting points, purity, and glass transition temperature (T_g) of pharmaceutical compounds. Qualitative and quantitative information can be gained on the physical and chemical changes of a sample. These changes are measured as a function of time and temperature.

DSC instruments present a closed system where the internal energy dU can not be created or destroyed. The system is governed by the First Law of Thermodynamics,

$$dU = dq + dw \tag{1.10}$$

where dq and dw are the heat transferred and the work done to the system. At constant pressure $(\ )_p$, the system experiences energy exchange with the environment in form of a change in enthalpy dH,

$$(dH)_p = (dU)_p + (PdV)_p \tag{1.11}$$

where PdV is the change in amount of work. As a approximation for solids and liquids, the change in volume can be neglected. If the system is also at zero net work, Equation 1.11 reduces to

$$(dH)_p = (dU)_p = (dq)_p \tag{1.12}$$

In this case, the heat flow (dq) in the DSC heating a sample is equal a change in enthalpy. A change in enthalpy is further related to the heat capacity Cp,

$$dq = dH = \int_{T_2}^{T_1} CpdT \tag{1.13}$$

where the increase in temperature (T_1 to T_2) is a function of the heat capacity of the system.

There are two types of instrumentation for DSC, a heat-flux and a power-compensated DSC (Figure 1.7). The samples S is heated in an aluminum pan together with an empty pan as the reference R. In case of heat-flux DSC, sample and reference are heated in a single furnace and the temperature difference between sample and reference is measured by heat-flux. In case of power-compensated DSC, sample and reference are heated by dual heaters and the difference in power consumption to keep sample and reference at the same temperature is measured. The power-compensated DSC has a higher sensitivity and allows analysis at higher temperatures.

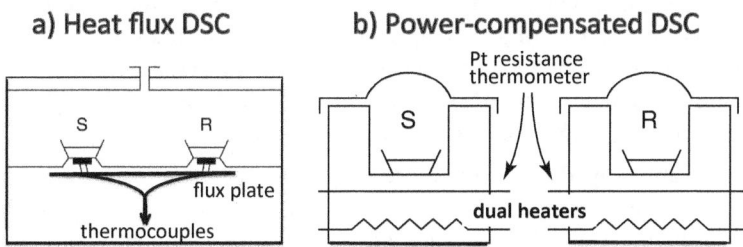

Figure 1.7: Schematic diagram of heat-flux (a) and power-compensated (b) DSC adapted from Clas et al. (1999).

Table 1.9: Enthalpic transitions observed by DSC (Clas et al., 1999).

Endothermic	Exothermic
Fusion	Crystallization
Vaporization	Condensation
Sublimation	Solidification
Desorbtion	Adsorption
—	Chemisorption
Desolvation	Solvation
Decomposition	Decomposition
Reduction	Oxidation
Degradation	Degradation
—	Curing of resins
Glass transition (baseline shift)	—
Relaxation of glass	—

The physical and chemical changes occurring in DSC measurements give thermal events in form of enthalpies. They may stem from diverse thermal reactions and they can be grouped into exothermic or endothermic transitions; an overview is given in Table 1.9.

1.3 Morphology

A crystalline solid may exhibit different crystal or particle shape although same polymorphic form is specified. The external shape of a crystalline compound is referred to as the crystal morphology. There are numerous crystal morphologies such as acicular, dendritic, cubic, fiber-like, plate-like, prismatic, rod-shaped, needle-like, as aggregates, or with crystal defects, to name a few (Rodrìguez-Hornedo et al., 2007). In a review on the solid-state properties of powders York (1983) showed that variations in crystal morphology influence the processing and product development. Morphological changes can effect particle size distribution, powder flow, mixing and agglomeration, dissolution, compression, and tablet hardness.

The crystal habit may also be engineered to improve the solubility and the dissolution rate of a drug. The bigger the surface area and the smaller the particle size, the better the wettability and the dissolution rate of the drug (Blagden et al., 2007).

1.3.1 Crystal Habit

The final crystal shape is given by several parameters. The polymorphic form is determined by the geometric arrangement of the molecules to a specific crystal lattice, the unit cell. Further, the unit cell can be described by crystal planes, the faces of a crystal. When a crystal grows, not all crystal faces grow the same and a distinct crystal habit is obtained.

Crystal faces can be described by the Miller indices h, k, and l. They are the intercept of the planes with the unit cell axes and they are given as reciprocals and even numbers. If a plane is parallel to a crystal axis the Miller index is zero. Describing a single plane the Miller indices are written in parenthesis (hkl), while describing a whole family of faces they are written in braces {hkl}. An example on how the faces can grow into different crystal habit is shown in Figure 1.8 (Rodrìguez-Hornedo et al., 2007).

1.3.2 Crystallization

When studying the crystal morphology of a drug it is further important to understand the process of crystallization. Crystallization starts with aggregation of molecules to prenucleation clusters. A nucleus of certain size existing during sufficient time may grow into a macroscopic crystal. According to Rodrìguez-Hornedo et al. (2007) the rate and mechanism of crystallization from liquid solution is influenced by a wide range of factors, namely:

- Solubility of the crystallizing compound
- Solvent
- Supersaturation of components participating the crystallization process

Morphology 23

- Rate at which a supersaturation is created, e.g., cooling or pH-change
- Temperature
- pH
- Soluble additives and impurities
- Reactivity of surfaces toward nucleation
- Diffusivity or viscosity
- Volume of solutions

Supersaturated states of drug or excipients may easily occur during pharmaceutical processing and drug delivery and thus facilitate crystallization. Furthermore, excipients can function as impurity and influence the habit of the crystal formed. A change in morphology could have a negative effect on processing and delivery (Rodrìguez-Hornedo et al., 2007).

Nucleation. The step of nucleation can occur over a homogeneous or over a surface-catalyzed mechanism. Homogeneous nucleation is the usual mechanism in small scales of less than 100 ml. The surface-catalyzed nucleation can occur either catalyzed by the surfaces of the crystallizing solutes (secondary nucleation) or as a so-called heterogeneous nucleation catalyzed by interfaces or surfaces other than the crystallizing solutes. Heterogeneous nucleation is of practical importance in drug formulations, as excipients may provide the surfaces or interfaces necessary to

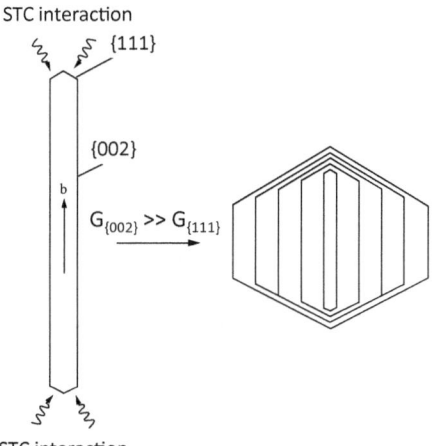

Figure 1.8: Crystal growth in favor of one crystal face leads to a distinct crystal habit: CBZ dihydrate shows much stronger growth from the {002} face than from the {111} faces if crystallized in presence of sodium taurocholate (STC); figure according to Rodrìguez-Hornedo et al. (2007).

promote nucleation. Furthermore, heterogeneous nucleation occurs at low driving forces and the favorable dissolution properties of a metastable drug may be lost because of fast nucleation of the stable form (Rodrìguez-Hornedo et al., 2007; Blagden et al., 2007).

Crystal Growth. If a nucleus overcomes a certain critical cluster size it is stable enough and can grow into macroscopic crystals. The process of crystal growth can be described by four stages: (1) the growth unit is transported to the impingement site, (2) it adsorbs, (3) the growth unit diffuses from the impingement site to the growth site, and (4) is incorporated into the space lattice to form the crystal structure. The mechanism of crystal growth can be volume-diffusion controlled or surface-integration controlled. Further, the surface-integration can occur as continuous crystal growth or by layers. The process of crystal growth in influenced by the general factors mentioned above. However, also the three dimensional structure and crystal defects can influence the growth rates and thus determine the interaction between surface and solution at a molecular level (Rodrìguez-Hornedo et al., 2007).

1.3.3 Analytical Techniques

Crystal morphology is best assessed by microscopical methods like scanning electron microscopy.

Scanning Electron Microscopy (SEM)

Electron microscopy uses electrons instead of light for the imaging, however, the same principles as for any light microscopy apply.

The Abbe's equation determines the resolving power as the minimum resolvable separation distance d_0 in [nm],

$$d_0 = \frac{0.61 \cdot \lambda}{n \sin \alpha} \quad (1.14)$$

where λ is the wavelength of the light, n the refractive index of the medium between sample and objective, and α the half-angle subtended by the objective at the sample (Schmidt, 2007).

The dual character of an electron to move as a particle or to radiate with a distinct wavelength was shown by Louis de Broglie (Equation 1.15),

$$\lambda = \frac{h}{mv} \quad (1.15)$$

where λ is the wavelength [nm], h the Planck's constant, m the mass of the electron, and v the velocity of the electron. This phenomenon is the basis for the electron microscopy, where on incidence of a primary electron beam a current of secondary electrons is reflected that can be collected for the imaging (Schmidt, 2007).

Morphology

The visualization of a sample at high resolution is of great interest to study crystal morphology. In pharmaceutical technology also the visualization of powder mixtures, granules, or tablet surfaces are essential to a deeper understanding of the sample behavior in drug development and drug delivery (Schmidt, 2007).

Powder Flow

Highly dependent on the particle morphology is the powder flow, an important parameter in pharmaceutical solids. Only powder with good flowability reaches uniform dosing in tablet press, by a good material flow from tablet hopper to the die. Flowability can be evaluated by the Hausner Ratio HR (Equation 1.16) and the Compressibility Index CI [%] also named Carr Index (Equation 1.17),

$$HR = \frac{\rho_t}{\rho_b} \quad (1.16)$$

$$CI = \frac{\rho_t - \rho_b}{\rho_t} \cdot 100 \quad (1.17)$$

where ρ_b is the bulk density and ρ_t is the tapped density in [g/cm^3]. The Hausner ratio shows a cut-off value of 1.25. Powder with a Hausner ratio below 1.25 indicates good flow, while poor flow is indicated by a Hausner ratios above 1.25. The compressibility index can classify powder flow according to Table 1.3.3 (USP 31, 2008).

Table 1.10: The classification of powder flow by the Compressibility Index (USP 31, 2008).

Compressibility Index [%]	Flow Character
≤ 10	excellent
11 – 15	good
16 – 20	fair
21 – 25	passable
26 – 31	poor
32 – 37	very poor
> 38	extremely poor

1.4 Solubility and Dissolution Rate

Solubility is defined as the concentration of drug dissolved per amount of solvent medium at the equilibrium. It is temperature and pressure dependent. In case of poorly soluble drugs a dissolution rate can assess the relative solubility of polymorphs much faster than over the time consuming equilibration process for the solubility assessment and with less medium. The *dissolution rate* is defined as the amount of solid drug dissolving per time unit under controlled liquid-solid interface, media condition, and temperature (Brittain and Grant, 1999). The Noyes-Whitney Equation (1.18) describes the process as a dissolution of a planar surface,

$$\frac{dC}{dt} = \frac{DS}{Vh}(C_s - C) \qquad (1.18)$$

where, C is the concentration of the solute in time t [mg/ml/min], D the diffusion coefficient, S the surface area [cm^2], V the volume of dissolution media [ml], h the thickness of the diffusion layer [mm], and C_s is the saturated concentration [mg/ml]. The dissolution is described as mass transfer through a stagnant film by diffusion, where the gradient is the most important driving force (Wesselingh and Frijlink, 2008). A schematic view of the diffusional dissolution process is shown in Figure 1.9.

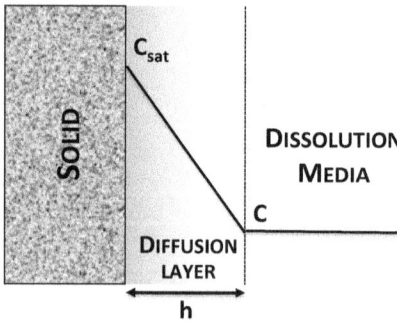

Figure 1.9: Dissolution principles of a solid drug by diffusion through a stagnant film (diffusion layer); schematic figure adapted from Wang and Flanagan (2008).

In perfect sink, the concentration in the dissolution media is much less than in saturated state ($C_{sat} \gg C_{sol}$). In experimental conditions, the volume of the dissolution media should be at least tree times as large as for the saturated state of the solid analyzed (Hanson and Gray, 2004a).

A common method to characterize pharmaceutical solids is by intrinsic dissolution rate (IDR). The mathematical expression describing IDR can be deducted from Equation (1.18). If $C_s \gg C$, and D/h is expressed as intrinsic dissolution constant k, Equation (1.18) can be written as,

$$\frac{dC}{dt} = \frac{S}{V} \cdot k \cdot C_s \quad (1.19)$$

and rearranged to Equation (1.20), the general expression for IDR (Kobayashi et al., 2000; Sethia and Squillante, 2004),

$$IDR = \frac{dC}{dt} \cdot \frac{V}{S} = k \cdot C_s \quad (1.20)$$

where ($\frac{dC}{dt}$) is the amount of drug dissolving over time, (S) the surface area of the compact, V the volume of the dissolution media, (k) the intrinsic dissolution constant, and (C_s) the solubility of the drug. Intrinsic dissolution curves are linear and the slope is the amount of drug dissolved per cm^2.

The IDR depends on the solid state properties of the drug, e.g., crystalline state, morphology, particle size, and particle surface area. Furthermore, experimental conditions such as disk rotation (hydrodynamics), fluid viscosity, dissolved gasses, temperature, and pH may influence the dissolution rate (Hanson and Gray, 2004b).

Under fixed experimental conditions IDR value is characteristic for a given substance. Compared to the equilibrium solubility IDR is a rate phenomenon and may therefore correlate better with the in vivo dissolution dynamics. IDR also presents an alternative measure to the equilibrium solubility to classify drugs, where an IDR < 0.1 mg/min/cm^2 classifies for low solubility (Yu et al., 2004; Zakeri-Milani et al., 2009). Lately, Tsinman et al. (2009) were able to determine IDR with sample quantities of 10'000-fold smaller than by the traditional rotating disk method using powder-based IDR measurement and bioexponential equations. Their findings promote IDR measurements as the method of choice to assess drug solubility in early drug development.

1.4.1 Disk Intrinsic Dissolution Rate (DIDR)

Intrinsic dissolution rate is measured from a constant surface area. The powder sample is compressed into a pellet and presented to the dissolution media in a disk allowing contact to one surface only. The settings for DIDR measurements can be in two modes, by a static disk and a paddle to stir the dissolution medium or by a rotating disk. In this PhD study a commercially available dissolution machine (Sotax*AT7smart*) was modified at its rotating basket unit to a rotating disk unit. Powder samples were compacted with controlled compaction parameters externally and placed into the sample holder (disk) after elastic recovery. The compacts were embedded with melted paraffin wax so only one surface was available to the dissolution medium (Figure 1.10).

DIDR was reported to increase directly proportional to the square root of the rotational speed, whereas the distance between disk and paddle has no effect (Yu et al., 2004). Further, the robustness of the values measured decreases with the rotation speed and with time (Mauger et al., 2003).

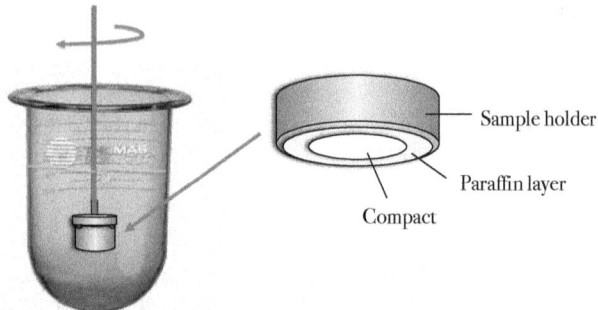

Figure 1.10: Schematic picture of a rotating disk apparatus to measure IDR. Modified rotating basket mode as developed within the PhD studies.

1.5 Carbamazepine (CBZ)

Carbamazepine, 5H-dibenz[b,f]azepine-5-carboxamide (Figure 1.11), and its antiepileptic activity was discovered by Schindler and Häfliger (1954). By now

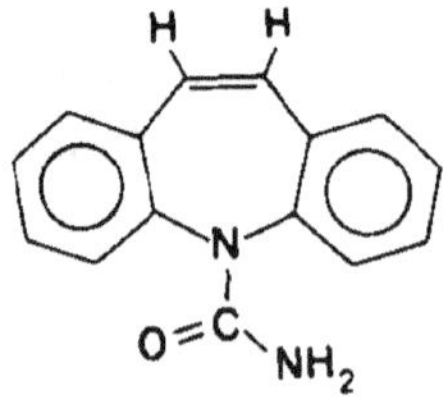

Figure 1.11: Carbamazepine (Himes et al., 1981).

Carbamazepine (CBZ) is a well established drug on the international market and several generics are available. CBZ is poorly soluble and with a dissolution controlled bioavailability, it is a class II according the Biopharmaceutics Classification System (BCS) and shows a narrow therapeutic index (Lindenberg et al., 2004). Furthermore, CBZ is a polymorphic drug and it exhibits at least four polymorphic forms, referring to p-monoclinic (form III), triclinic (form I), c-monoclinic (form IV), and trigonal (form II) crystal lattice. In literature CBZ polymorphs show dissimilar naming. In this study CBZ polymorphs are named according to Grzesiak et al. (2003) to avoid confusion. The different packing of the four polymorphic forms is shown in Figure 1.12. The CBZ molecules arrange to an anti-carboxamide dimer motif and the four polymorphic forms are then formed by a distinct packing of these dimers (Rodrìguez-Sponga et al., 2004).

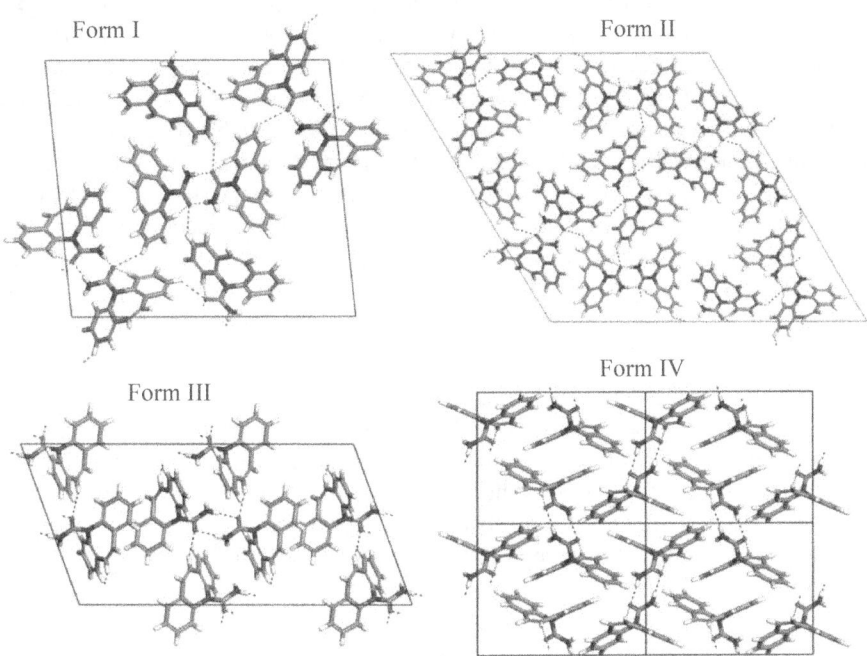

Figure 1.12: Packing diagrams of carbamazepine polymorphs form I, II, III, and IV (Rodrìguez-Sponga et al., 2004).

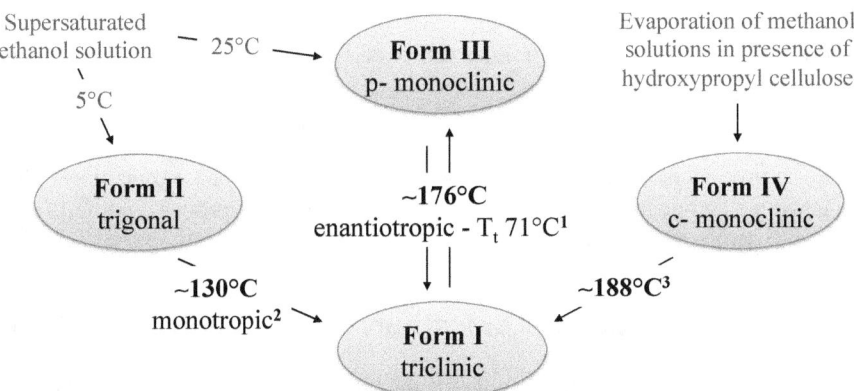

Figure 1.13: Schematic overview of thermal transition between CBZ polymorphs, adapted from Grzesiak et al. (2003); [1]Behme and Brooke (1991), [2]Edwards et al. (2001), and [3]Grzesiak et al. (2003).

The thermal transitions between the CBZ polymorphs is shown in Figure 1.13. CBZ form II, III, and IV can all transform to form I (Grzesiak et al., 2003). Form III and I show an enantiotropic relation with a transition temperature (T_t) of 71°C, above this temperature form I is the stable form (Behme and Brooke, 1991). At room temperature CBZ form III is the thermodynamically most stable form and it is also the form commercially available. The order of thermal stability is reported as form III > form I > form IV > form II. The order complies with the rule of density, with 1.34, 1.31, 1.27, and 1.24 g/cm^3 for form III, I, IV, and II, respectively (Grzesiak et al., 2003). Interestingly, form IV it is not the thermodynamically least stable form, although it was the last CBZ polymorph discovered (Lang et al., 2002). An explanation may be the crystallization procedure, form IV has only been reported to crystallize in presence of the polymer hydroxypropyl cellulose (Florence et al., 2006).

CBZ can further exist as solvate with acetone or as cocrystal with saccharin, nicotinamide, acetic acid, or with 5-nitroisophthalic acid (Fleischman et al., 2003; Rodrìguez-Sponga et al., 2004). The most common solvate is with two molecules of water, the CBZ dihydrate and it is the thermodynamically most stable form in aqueous solutions and at high relative humidity (Figure 1.14).

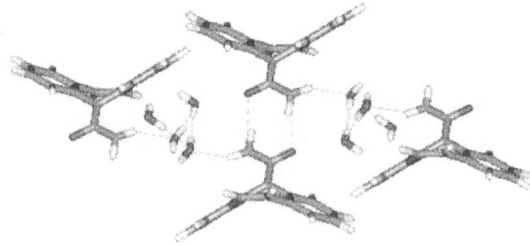

Figure 1.14: The geometrical arrangement of carbamazepine dihydrate (Rodrìguez-Sponga et al., 2004).

CBZ is a prominent model drug to study polymorphism and phase transitions. CBZ is also selected because of its low solubility to investigate process methods as to their ability to enhance solubility.

1.5.1 Transformation of Anhydrous CBZ to its Dihydrate Form

The transformation process best studied is the one from anhydrous CBZ to the dihydrate form. Laine et al. (1984) described the transformation as solution-mediated process and the formation of dihydrate crystals by whisker growth. The dehydration mechanism of CBZ was described by Han and Suryanarayanan (1998), they reported the transformation to occur phase boundary or nucleation controlled. Anhydrous

CBZ (form III) and CBZ dihydrate were reported to coexist in the water activity range of 0.601–0.641 at 20 °C (Qu et al., 2006; Li et al., 2008).

Several attempts have been made to control the transformation of CBZ anhydrous to dihydrate, mostly by inhibiting the nucleation and the formation of CBZ diyhdrate crystals. Dihydrate formation is strongly inhibited in solutions (1–4% w/v) of hydroxypropyl methylcellulose (HPMC), hydroxypropyl cellulose (HPC), methyl cellulose (MC), or poly-vinyl pyrrolidone (PVP), whereas solutions of polyvinyl alcohol (PVA), polyethylene glycol (PEG), sodium carboxymethyl cellulose (CMC), and polysorbate 80 (Tween® 80) have only weak inhibitory activity (Gift et al., 2008; Tian et al., 2006; EL-Massik et al., 2006; Otsuka et al., 2000; Luhtala, 1992). In contrast, surfactants, such as sodium laurylsulphate (SLS) and sodium taurocholate (STC) promote the dihydrate formation during the dissolution test (Rodrìguez-Hornedo and Murphy, 2004).

Few authors studied the effect on transformation by including the excipient directly into the tablet formulation. Katzhendler et al. (1998, 2000) analyzed the gel layer of matrix tablets containing HPMC and later also egg albumin during the dissolution process. HPMC inhibits the transformation to the CBZ dihydrate form, however, HPMC also participates in its crystallization process and induces amorphous CBZ. Egg albumin also inhibits the transformation, though with a dose dependence. Salameh and Taylor (2006) studied the transformation exposing physical mixtures of anhydrous CBZ or theophylline with mannitol, microcrystalline cellulose (MCC), PVP K12, or PVP K90 to various relative humidity. MCC has only minimal effect on both drugs, mannitol enhances dehydration, whereas PVP K12 and PVP K90 showed contradictory results. The authors concluded, that the effect of excipients on hydrate forming drugs depends on a multitude of factors and knowledge is still limited. Recently, Schulz et al. (2010) successfully adsorbed CBZ to crospovidone (CrosPVP) to prevent recrystallization to CBZ dihydrate during the solvent disposition. However, transformation is prevented only in small drug loads ($\leq 9.1\%$). Also a surface modification of the CBZ crystals seems to inhibit dihydrate formation. Otsuka et al. (2003) adsorbed n-butanol on CBZ crystals of triclinic form (form I). Compared to the untreated crystals, surface-modified crystals were more stable towards the influence of moisture and showed a faster dissolution rate.

Transformation of anhydrous CBZ (form III) to its dihydrate has also been studied at the level of a single crystal (Tian et al., 2006). Carbamazepine crystals transform first at the sites of defect and the crystal faces transform with different speed. The smaller faces of the CBZ crystal (010 and 001) grow the most dihydrate crystals, whereas the largest face (100) grows much less. Larger crystal faces interact less with polar solvents such as water and by dissolving at a slower rate, less dihydrate crystals can be formed. CBZ dihydrate crystallized from water are of characteristic needle shape, the crystal grows fastest along the direction of (010), the crystal face $\{111\}$. Also in presence of SLS the needle-shaped habit is grown. In contrast, presence of STC inhibits the growth along the $\{111\}$ direction and the CBZ dihydrate crystal grows more along the direction of the $\{200\}$ and $\{002\}$ faces (see also Figure 1.8 in Section 1.3.1). (Rodrìguez-Hornedo and Murphy, 2004)

1.5.2 Mechanical Properties of CBZ

CBZ is poorly compressible, and its compactibility depends on the polymorphic form. Roberts and Rowe (1996) compared CBZ form III and I in Young's modulus and yield stress. The thermodynamically more stable form has a higher true density, higher yield stress, higher Young's modulus, it is less compressible and it is therefore more difficult to form compacts. In a later study they were able to predict the mechanical properties of CBZ by an atom-atom potential model applied to lattice dynamics. However, this was only possible if the crystal morphology was also considered (Roberts et al., 2000). The compaction of CBZ shows high amount of elastic behavior and also compaction-induced disorders and polymorphic transition can occur (Koivisto et al., 2006). CBZ dihydrate shows the best compactibility, however, it is not stable under compression (Lefebvre et al., 1986).

1.5.3 CBZ and Analytical Methods

Analysis methods to study CBZ polymorphs and their transformations have evolved fast over the last decades. To verify polymorphic form, X-ray powder diffraction (XRPD) is the method of choice. Figure 1.15 shows the typical patterns for the CBZ polymorphs. Also the thermal analysis by differential scanning calorimetry (DSC) is a prominent analysis tool to study polymorphic transition. Figure 1.16 shows the typical thermograms of CBZ polymorphs and of amorphous CBZ. Upon heating CBZ form II, III, IV, and amorphous all transform to form I.

Edwards et al. (2001) proposed the use of time and temperature resolved X-ray diffraction techniques to analyze polymorphic transition of CBZ. However, to reduce the effect of particle size, morphology, and preferred orientation limiting the analysis by X-ray diffraction Tian et al. (2007) suggested the use of Raman spectroscopy. They successfully applied in situ Raman spectroscopy to follow CBZ anhydrous to dihydrate transformation quantitatively. Raman spectroscopy also seems more discriminative towards CBZ polymorphs and dihydrate than the analysis by near infrared spectroscopy (NIR) (Kogermann et al., 2008). Earlier, O'Brien et al. (2004) successfully applied in situ FT-Raman spectroscopy to study the kinetics of thermal transitions in CBZ. Another in-situ technique to analyze CBZ transformation was reported by Brittain (2004), they were able to follow the solution-mediated transformation of CBZ anhydrous (form III) to its dihydrate using fluorescence studies and they reported first-order reaction kinetics for the transformation process. Following the thermal transformation of CBZ form III to I by time-resolved terahertz spectroscopy Zeitler et al. (2007) could quantitatively follow its solid-gas-solid transitions.

CBZ polymorphs can also be distinguished by their intrinsic dissolution rate (IDR). However, IDR values strongly depend on the measurement parameters. For CBZ IDR values are reported in the range of 0.024–0.077 mg/min/cm^2 (Kobayashi et al., 2000; Yu et al., 2004; Zakeri-Milani et al., 2009; Šehić et al., 2010). CBZ dihydrate shows IDR values of up to 2.5 times lower than for anhydrous CBZ (Ono et al., 2002). The IDR analysis is in water or in aqueous solutions, therefore CBZ anhydrous transforms to the slower dissolving dihydrate form. This transformation is visible in IDR profiles as a change in slope, also named transformation point (Kobayashi et al., 2000; Šehić et al., 2010). Kobayashi et al. (2000) found that the transformation depends on the polymorphic form. Šehić et al. (2010) reported different transformation points within CBZ samples of same polymorphic form but of different commercial suppliers. The difference also continued to exist in tablet formulations using Ludipress® as tablet filler.

The dissolution test described in the USP monograph for CBZ immediate release tablets requires a dissolution medium of 1% SLS in water. This can lead to inhomogeneous dissolution behavior of CBZ tablets because SLS enhances transformation of CBZ to its dihydrate. EL-Massik et al. (2006) proposed an alternative composition of the dissolution medium containing 0.5% SLS and 0.01% MC. This composition removes induction of transformation without compromising the solubility enhancement by SLS, it provides suitable sink for up to 400 mg CBZ, and it is robust to various dosing strength of same CBZ formulation.

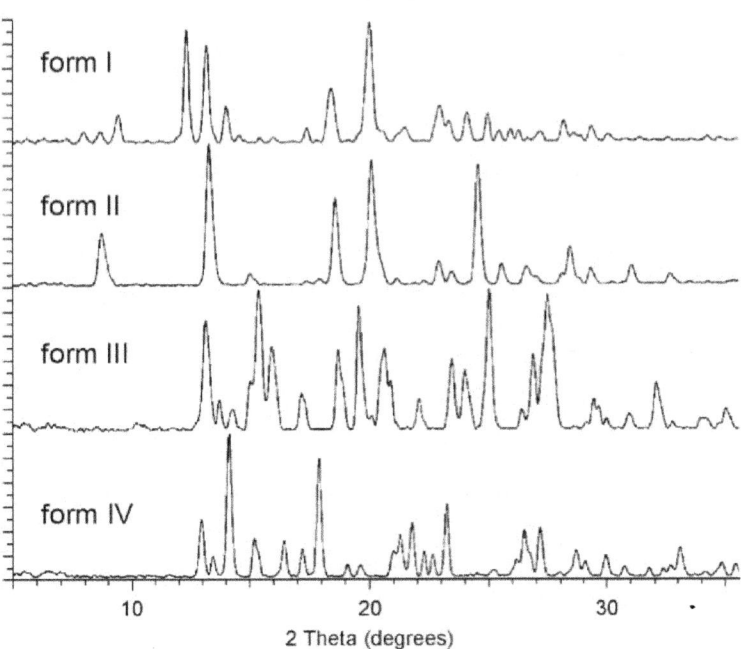

Figure 1.15: Diffractograms of CBZ polymorphic forms I–IV (Rodrìguez-Sponga et al., 2004).

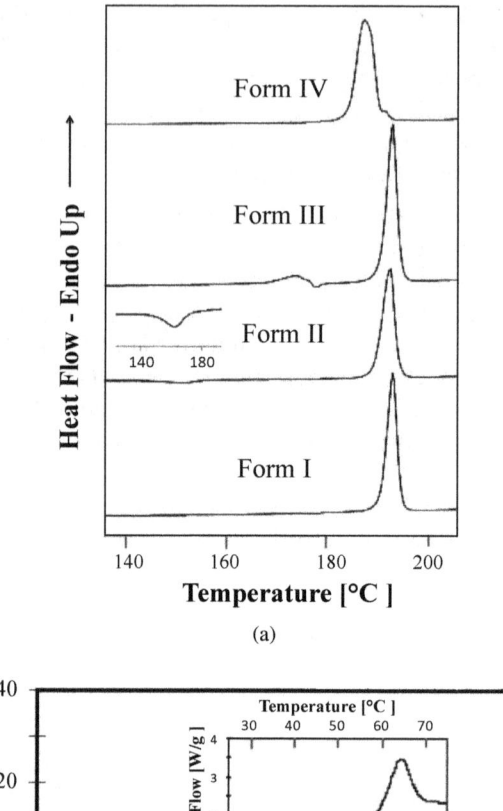

Figure 1.16: DSC of CBZ polymorphic forms I–IV (a) and amorphous CBZ (b) upon heating (Rodrìguez-Sponga et al., 2004; Li et al., 2008).

1.5.4 Techniques to Enhance CBZ Dissolution Rate

Several techniques are reported to enhance the dissolution rate of CBZ, e.g., crystal engineering, formation of water-soluble complexes, solid dispersion with surfactants, formulation of a metastable or amorphous state, or reducing particle size by micronization (Javadzadeh et al., 2007). A process method may include more than one aspect enhancing the dissolution rate, a solid dispersion with surfactants may enhance dissolution also by the process-induced particle size reduction and by a changed solid state.

Crystal Engineering. The dissolution rate of CBZ can be enhanced by a different crystal habit and smaller crystals. Bolourtchian et al. (2001) reported an improved dissolution rate and compactibility of CBZ crystals prepared by different solvent techniques. Crystallization from ethanol or acetone resulted in polyhedral and thin plate-like crystals, respectively. Using a watering-out method needle-shaped crystals were obtained. Nokhodchi et al. (2005) studied recrystallization of CBZ from alcohol solutions in presence of the additives PEG 4000, PVP K30, and Tween 80. The obtained crystals were of different crystal habit but of same polymorphic form (CBZ form III). However, only the recrystallization in presence of 1% PVP K30 resulted in more block-shaped crystals with improved dissolution rate and tensile strength.

Also the formation of cocrystals can provide a possibility to enhance the dissolution rate of CBZ (Rodrìguez-Sponga et al., 2004). Hickey et al. (2007) reported favorable dissolution properties of the CBZ:saccharin cocrystals in table formulation.

Solid Dispersions. Attia and Habib (1985) studied the effect of sugars (i.e., mannitol, lactose, and galactose), PEG 6000, and PVP on the dissolution rate of CBZ using solid dispersions. Highest dissolution rates were obtained from the CBZ–PVP coprecipitate. Zerrouk et al. (2001) included PEG 6000 in CBZ solid dispersions by a melt technique, however, CBZ transformed to its form II during the processing. The solid dispersion by coprecipitates or melts allow particle size reduction to a molecular level. Nonetheless, the solid dispersion method has to be selected and controlled carefully because of possible phase transitions induced by the process. El-Zein et al. (1998) studied different carriers to improve dissolution of CBZ. Solid dispersion with the hydroxypropyl-β-cyclodextrin showed the best dissolution rate and in-vivo absorption, the absorption was also higher than from the original product, the Tegretol suspension. Nair et al. (2002) prepared solid dispersions from methanol solutions. They reported amorphous CBZ in solid dispersion prepared in presence of PVP K30, whereas solid dispersions prepared in presence of high amounts of PEG 4000 (PEG:CBZ, 6:4) resulted in CBZ form I. Both solid dispersions showed an enhanced dissolution rate. Compared to untreated CBZ, the formation of dihydrate was much higher because of the change to a less stable solid state. A solvent free method was applied by Moneghini et al. (2001) using supercritical carbon dioxide to obtain CBZ solid dispersions. The processed CBZ did not change polymorphic form (form III), but dissolution rate was enhanced due to a smaller particle size. Also Gosselin et al. (2003) successfully micronized CBZ without changing the polymorphic form. They applied a method using rapid expansion of supercritical carbon dioxide solutions (RESS) for the micronization. CBZ form III was obtained keeping the process below 40°C and below 240 bar. Ugaonkar et al. (2007) prepared amorphous CBZ by a solid dispersion method using near-supercritical carbon dioxide (n-scCO$_2$) and the polymer PVP. A metastable or amorphous solid has the advantage of a higher solubility and dissolution rate. The disadvantage is that a pure phase should be obtained and kinetically stabilized.

Liquisolid Technique. Javadzadeh et al. (2007) successfully applied the liquisolid technique to improve dissolution properties of CBZ. MCC was used as carrier and aerosil as coating material, the liquid load was up to 0.6 and the excipient ratio was 15 or 20. Presence of PVP could increase the dissolution rate and by including PEG 35000 an increased liquid load was achieved without decreasing powder flowability. No polymorphic changes were detected by XRPD or DSC measurements after the liquisolid processing.

Adsorption on Nanopores. Ambrogi et al. (2008) reported increased drug release of CBZ adsorbed to mesoporous silicate MCM-41. CBZ loading was 14% and in noncrystalline state. Moreover, the system was not sensitive to moisture and elevated temperature.

1.5.5 Irregular Dissolution, Bioinequivalence, and Clinical Failures in CBZ Tablets

CBZ tablet formulations show a history of irregular dissolution behavior, bioinequivalence, and clinical failures. A large difference in bioavailability and in vitro dissolution rate among CBZ tablets was reported by Meyer et al. (1992). The analyzed CBZ tablets were of different lots withdrawn from the market due to several reports of clinical failures. In a later study, Meyer et al. (1998) simulated steady-states of the various CBZ tablets. However, they could not show clinical relevance by these simulations, despite an observed bioinequivalence of single doses. Bioinequivalence of various marketed CBZ tablets are also shown by Jung et al. (1997) and Lake et al. (1999), where the area under the plasma level curve (AUC) and maximum plasma level (C_{max}) clearly differ. Wang et al. (1993) provided a possible reason for the clinical failures of CBZ tablets. They reported an extensive effect of high humidity and temperature on disintegration and on drug release of CBZ tablets. In a multinational survey on in vitro dissolution, Davidson (1995) showed high variability in commercial CBZ tablets even among the same brand. The variability was highest after 15 min dissolution testing, where 45–75% drug must be released according to the USP monograph for CBZ tablets. Almost ten years later, marketed CBZ tablets still vary widely in their dissolution behavior Mittapalli et al. (2008).

1.5.6 CBZ Tablets Registered in Switzerland

By the beginning of 2011 the following CBZ tablet formulations are registered on the Swiss market (Kompendium, 2011):

- **Tegretol®** by Novartis Pharma (original); tablets and CR divitabs of 200 mg and 400 mg.
- **Carsol®** by Sandoz; CR tablets of 200 mg and 400 mg.
- **Timonil® /-retard** by Desitin; tablets of 200 mg, 300 mg, 400 mg, and 600 mg.
- **Neurotop® retard** by Orion Pharma; Tablets of 300 mg and 600 mg. Excipients known to the public: Eudragit RS PO and L 30 D, highly disperse silicium dioxide, magnesium stearate, talcum, sodium carboxymethylstarch, and MCC.

1.6 Formulation Studies

"....it is a medicine, not a molecule, that we are giving to the patient." (Brown, 2005).

After careful preformulation studies an active pharmaceutical ingredient (API) is formulated into an appropriate dosage form – the medicine. The most popular dosage form is the oral tablet formulation. To formulate an API into a tablet of satisfactory quality the solid state of the API must have the following three essential attributes by itself or in a formulation with selected excipients. (1) a rapid and reproducible flow into the die of the tablet press to ensure uniformity in tablet weight and content, (2) the particles must cohere under compression and that coherence should remain after the compression force is removed, and (3) the formed compact should be removed from the tablet press without damage of the compact or the tablet press. APIs rarely possess all three attributes, thus preliminary treatment are paramount. The most important methods to incorporate these attributes to a tablet formulation are the wet granulation, dry granulation, and the direct compaction methods (Armstrong, 2007a).

1.7 Direct Compaction

To manufacture tablets of thermolabile and moisture-sensitive drugs such as CBZ direct compaction is the technique of choice. Direct compaction is sometimes also referred to as direct compression. However, *compressibility* is defined as the ability of a powder to densify under pressure, whereas *compactibility* is the ability to form a tablet of specific hardness (Alderborn and Nyström, 1996). Therefore, direct compaction is the term selected referring to the process method within this work.

Direct compaction provides the following advantages over wet or dry granulation (Jivraj et al., 2000):

- Requires fewer operation steps, resulting in shorter processing time and lower energy consumption.
- Fewer stability issues for actives that are sensitive to heat or moisture.
- Fewer excipients may be needed in a direct compaction formula, see Table 1.11.

Also the following disadvantages should be considered (Jivraj et al., 2000):

- Segregation issues – however, reduced by matching the particle size and density of the API with the excipients.
- Low bulk density, poor flowability and compactibility of the drug can jeopardize direct compaction, especially in low potency drugs (Carlin, 2008).
- Static charges may develop on the drug particles or excipients during mixing and thereby lead to agglomeration of particles resulting in poor mixing.
- For high potency drugs with a dosing of less than 1% content uniformity direct compaction is a challenge and with a dosing of less than 0.1% other processes such as the adsorption of the drug on carriers cannot be avoided (Carlin, 2008).

In direct compacted tablets the physicochemical properties of the drug dominates the final tablet properties. At low drug loads, however, the effect of the excipients dominate and variations in their physicochemical properties could critically influence the tablet performance. The excipients for direct compaction have to be selected with care. There are many specialities available for direct compaction and in most cases, these are common materials that are physically modified to provide better flow and compactibility characteristics (Carlin, 2008).

1.8 Choice of Excipients for Direct Compaction

The choice of excipients is critical for a successful tablet production by direct compaction. They should be specified for particle size, bulk density, flow, and compactibility. Most excipients for direct compaction are preprocessed. A tablet formula for direct compaction usually consists of the API and excipients functioning as filler-binder, disintegrant, and lubricant (Carlin, 2008). Table 1.11 gives an overview of the excipients and their concentrations in a tablet formula for direct compaction.

Choice of Excipients for Direct Compaction 39

Table 1.11: General tablet formula for direct compaction (Carlin, 2008).

Active pharmaceutical ingredient (API)	0.1–99%
Filler-binder (dependent on API loading and compactibility)	1.0–99%
Disintegrant	0.5–2%
Lubricant	0.5–2%

1.8.1 Filler-binder

The filler-binder is the most important excipient in direct compacted tablets, it often serves as the tablet matrix or vehicle. Filler-binder include a range of excipients from high compactibility and dilution capacity (e.g., MCC) to tablet fillers of low dilution capacity (e.g., spray-dried lactose). The dilution capacity of an excipient is its ability to carry poor compactible drug.

There are several factors to consider when selecting the appropriate filler-binder (Carlin, 2008):

- good compactibility
- good flowability
- bulk density
- particle size and distribution
- dilution capacity
- moisture content, water activity (free or bound water), hygroscopicity, effect of moisture on compactibility
- compatibility with API and stability in the final product
- solubility (influence on dissolution)
- physiological acceptability such as toxicity, osmotic effect, and taste
- cost and availability, security of supply
- regulatory acceptability (pharmacopoeial status, FDA's GRAS list of excipients; GRAS = generally recognized as safe)

Filler-binders can be classified according to their chemical similarity, Figure 1.12 gives an overview.

Microcrystalline Cellulose (MCC)

MCC is the most prominent filler-binder, it is highly compactible and can therefore dilute a poorly compactible drug. MCC is the only true binder, compact hardness can increase significantly by adding only small amounts (3–5% MCC). MCC is hygroscopic and practically insoluble in water. There are a number of grades

Table 1.12: Filler-binders suitable for direct compaction listed according to their chemical constitution (Bolhuis and Armstrong, 2006).

Cellulose	Microcrystalline Cellulose (MCC), Powdered Cellulose
Starch/Starch derivatives	Native Starches, Compressible Starch, Modified Starches
Inorganic salts	Dicalcium Phosphates, Tricalcium Phosphate, Calcium Sulfate Dihydrate
Polyols	Sorbitol, Mannitol
Lactose	α–Lactose Monohydrate, Anhydrous α–Lactose, Anhydrous β–Lactose Spray-Dried Lactose, Agglomerated Lactose
Other Sugars	Compressible Sugar, Dextrose, Dextrate
Co-processed Products	Ludipress®, Cellactose®, Pharmatose® DCL 40

available that are suitable for direct compaction, they differ in compactibility, particle size, and moisture content. The grades PH 102 and PH 101 are most common, they are of partially spray-dried MCC and particle size is around 100 μm and 50 μm, respectively (Rowe et al., 2006).

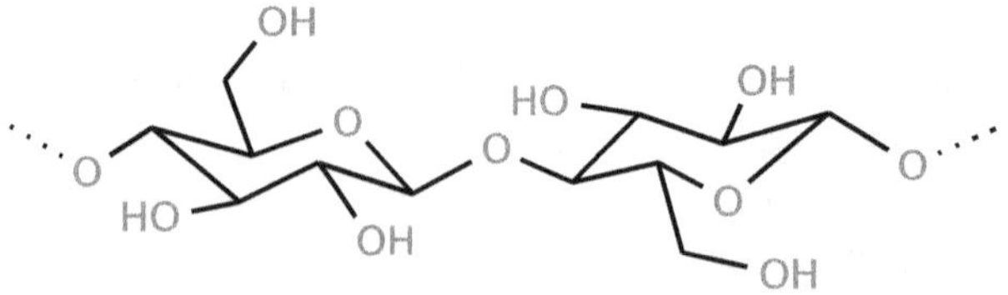

Figure 1.17: Repeat dimer unit of cellulose, linear chains of β(1–4)-linked-D-glucopyranosyl units (Carlin, 2008).

MCC is gained from natural α-wood cellulose by acid hydrolysis (Rowe et al., 2006). The smallest unit of cellulose is shown in Figure 1.17. Cellulose is a β(1–4)-linked polymer of glucose, two D-glucopyranosyl molecules build the repeat dimer of a linear chain (Carlin, 2008).

Mannitol

Mannitol belongs to the polyols and is an isomer of sorbitol (Figure 1.18). Direct compressible forms of several particle sizes are available (Armstrong, 2007a). Mannitol is non hygroscopic, water soluble, and exhibits three polymorphic forms (Rowe et al., 2006). Form III shows the best compactibility and it is of high

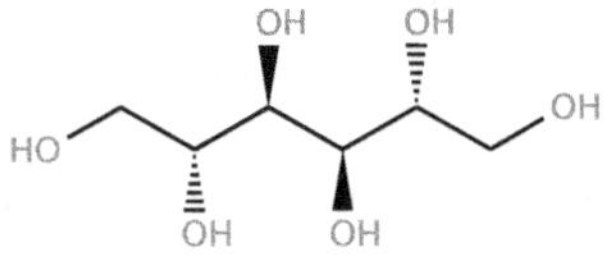

Figure 1.18: Structural formula of D-Mannitol (Rowe et al., 2006).

kinetic stability, however, form I is the thermodynamically most stable form at room temperature (Burger et al., 2000). Mannitol is a favored excipient in chewable tablets and lozenges because it shows a negative heat of solution and provides a cooling effect (Armstrong, 2007a).

Hydroxypropyl Cellulose (HPC)

HPC can express different functions depending on the degree of substitution and on the particle size. Alvarez-Lorenzo et al. (2000) evaluated low-substituted HPC (L-HPC) as filler-binders for direct compaction. Whereas HPC with a medium or high degree of substitution are suitable for matrix tablets, low-substituted HPC can be proposed as excipients for conventional tablets. Depending on the particle size L-HPC is effective as tablet binder or as tablet disintegrant. As fine powders, L-HPC shows low viscosity and sustained release. Drug release can also be regulated by the amount of L-HPC in the formulation (Kawashima et al., 1993). The compaction behavior of HPC has been studied by Picker-Freyer and Dürig (2007). They reported viscoelastic compaction behavior of HPC with increased compactibility and plasticity for lower degree of substitution and smaller particle size. These grades are therefore suitable for the use in direct compaction.

1.8.2 Disintegrant

A physically strong tablet can slow down the dissolution of the API, therefore tablet formulations often include a disintegrant. An excipient acting as a disintegrant disrupts the tablet structure and leads to fragmentation when it comes into contact with water; it facilitates the dissolution by exposing a larger surface area to the dissolution media (Armstrong, 2007b). Traditional disintegrants are the native and modified starches. They are poorly compactible and of low efficiency. Partially pregelatinized starches can be used as a disintegrant in formulations for direct compaction at 10–15% w/w (Moreton, 2008).

Superdisintegrants are much more effective and can be used at lower concentrations and they are thus better suited for tablets prepared by direct compaction. The most prominent superdisintegrants are croscarmellose (AcDiSol®), sodium starch glycolate (Eplotab®, Primogel®), and crospovidone (Polyplastone® XL). They are effective at levels of 0.5–4% (Carlin, 2008). Tablet disintegration occurs through different mechanisms. Sodium starch glycolate mainly acts as disintegrant through swelling, whereas the fiber-structured croscarmellose sodium further acts through energy recovering of elastic deformation and wicking (Moreton, 2008). The disintegrant activity of crospovidone is believed to occur mainly through the water wicking, the ability to draw water into the porous network of the tablet. The disintegration activity is not fully understood and several mechanisms may be needed to explain the complex behavior (Augsburger et al., 2002). *Crospovidone* is

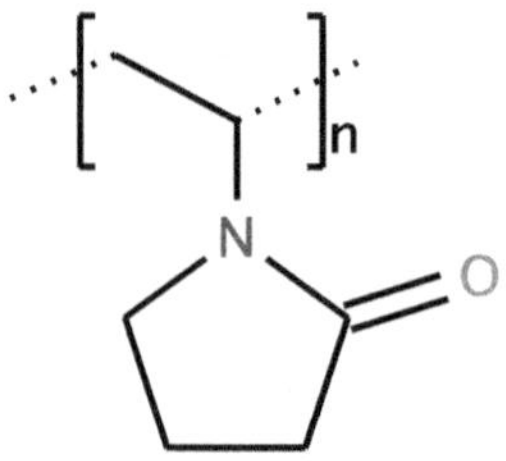

Figure 1.19: Molecular unit of polyvinylpyrrolidone (PVP) and its cross-linked form Crospovidone (Rowe et al., 2006).

cross-linked polyvinylpyrrolidone (Figure 1.19), it is hygroscopic and practically insoluble in water. The particles are of porous morphology and irregular shape, and the use of bigger particle size results in faster disintegrating tablets. For immediate release tablets the recommended level of incorporation is 2–5% (Rowe et al., 2006).

MCC. Tablets of pure MCC self-disintegrate in aqueous media when prepared by direct compaction. For an adequate disintegration, MCC levels of at least 20% are required and in case of hydrophobic API no sufficient disintegration may be obtained. The disintegration activity of MCC is believed to occur mainly through a combination of wicking and disruption of particle-particle bonds by the presence of water (Moreton, 2008).

L-HPC can provide another option of disintegrant in formulations for direct compaction. The fibrous and coarse grades are the ones used as disintegrant at levels of 2–10% (Moreton, 2008; Kawashima et al., 1993).

1.8.3 Lubricant

The main role of lubricants is to decrease the friction in tableting at the interface between tablet surface and die wall during ejection and to reduce the wear on punches and dies. Hydrophobic lubricants also decrease the interparticulate friction and thus enhance the powder flow. Furthermore, they are anti-adherent and can thereby prevent sticking to punch faces (Armstrong, 2008). Magnesium stearate is the most widely used lubricant and concentrations of 0.25–5% are used (Rowe et al., 2006). As a hydrophobic lubricant magnesium stearate can cause problems. In direct compaction with plastically deforming materials a thin film of magnesium stearate can cover the surface and interfere with the bonding. The result are weaker tablets with higher friability and because of the hydrophobic film with slower dissolution. The softening and hydrophobic effects can be minimized by adding magnesium stearate as the last step to the blend and by a very short mixing time of 2–5 min. Other lubricants may be required in some tablet formulation for direct compaction. Alternative lubricants are stearic acid, hydrogenated vegetable oil, or sodium stearyl fumarate (Carlin, 2008).

1.9 Analysis of Tablet Properties

A tablet formulation should deliver the active ingredient in an accurate and reproducible manner. Three general test categories ascertain the product quality. The first category includes tests on the nature of the API such as identity, quantity, and impurity and on the product properties such as tablet hardness and friability. In the second category bioavailability is established by in vitro and in vivo tests. The third category include tests on the product stability (Qureshi, 2007).

1.9.1 Tablet Hardness

Tablets experience a series of stresses during manufacturing, packing, shipping, and dispensing and the tablets should exhibit sufficient mechanical strength (Qureshi, 2007). Tablet strength can be assessed by the force required to fracture a tablet in a defined direction, the crushing strength. The physical strength of a tablet also depends on its dimensions and tensile strength provides a measure that includes tablet dimension. Tensile strength σ [MPa] of a tablet is given by the Equation (1.21),

$$\sigma = \frac{2F}{\pi dt} \qquad (1.21)$$

where F is the force [N] needed to fracture the tablet by a hardness tester, d the diameter [mm], and t the thickness [mm] of the cylindrical tablet (Armstrong, 2007b).

1.9.2 Tablet Friability

Tablet friability provides another measure to estimate the mechanical strength of the tablet. The abrasion of uncoated tablets is measured by subjecting them to controlled mechanical shock and attrition. The apparatuses to test friability are drums of specific diameter, width, and with a curved ramp over which the tablets role, slide, and fall at each turn. The standard method is performed with tablet of a total weight around 6.5 g, the drum is rotated at 25 rpm and with a total of 100 turns. The abrasion is measured as the loss in weight and should be less than 1% to pass the friability test (Qureshi, 2007).

1.9.3 Tablet Porosity

Tablet porosity is an important parameter in formulation studies. Although robust tablets of strongly coherent particles are essential, their low porosity can be an impediment to dissolution. The surface area exposed to the dissolution media increases with tablet porosity. Porosity (ϵ) can be calculated by the Equation (3.3) based on tablet volume,

$$\epsilon = (1 - \frac{m}{V_t \rho_t})100 \tag{1.22}$$

where m is the tablet mass, V_t the tablet volume, and ρ_t the true density of the powder blends (Armstrong, 2007b). Mercury intrusion porosimetry provides a further option to assess tablet porosity (Augsburger and Zellhofer, 2007).

1.9.4 Disintegration Testing

The disintegration test is a method of quality-assurance for conventional tablets. There are pharmacopoeial standards on test conditions, settings and outcome. The apparatus for the disintegration measurement consists of a basket rack with six tubes open at the top and confined by a 10-mesh screen at the bottom. One tablet is placed per tube and the basket rack is moved vertically at specific rate in aqueous media at 37 $\pm$ 2 °C. The test measures the time all six tablet need to disintegrate into aggregates and/or fine particles. Disintegration test does not necessarily correlate with the dissolution rate, therefore dissolution testing is often favored over disintegration (Qureshi, 2007; USP 31, 2008).

1.9.5 In Vitro Dissolution Testing

The dissolution can be measured by four general methods described by USP 31 (2008). The most prominent dissolution methods are the basket and the paddle method, the USP Apparatuses 1 and 2 (Figure 1.20). Dissolution can also be tested by the reciprocating cylinder method or in flow-through cell (Figure 1.21), the USP Apparatuses 3 and 4. In basket, paddle, and reciprocating cylinder method the dissolution media is of limited volume, while the flow-through cells allow measuring in an open system, the drug release is studied under continuous flow of fresh dissolution medium. This method is especially useful for analysis including pH changes and for analysis of poorly soluble drugs, the open system provides infinite sink. The flow rate is the critical parameter for the dissolution rate, typically a flow rate of 16 mL/min is applied (Hanson and Gray, 2004b). Also the size of the dosage form, its orientation within the dissolution cell, the cell size, and the gases dissolved in the media effect the flow of the media and thereby the dissolution rate. In contrast, deviations in temperature and conductivity of the medium have less impact on the dissolution (Cammarn and Sakr, 2000; Jørgensen and Jacobsen, 1992). The flow-through method can further be used to study intrinsic dissolution, Peltonena et al. (2003) reported a good correlation between the dissolution profiles obtained by channel-flow and intrinsic dissolution method.

Generally, the dosage form has to be placed with absolute care in dissolution measurements. Depending on the physical location, the dosage form experiences different shear strain environment, which is directly linked to different dissolution rate (Hanson and Gray, 2004b). Kukura et al. (2004) studied the paddle method using computational fluid dynamics. The dissolution vessel shows a heterogeneous

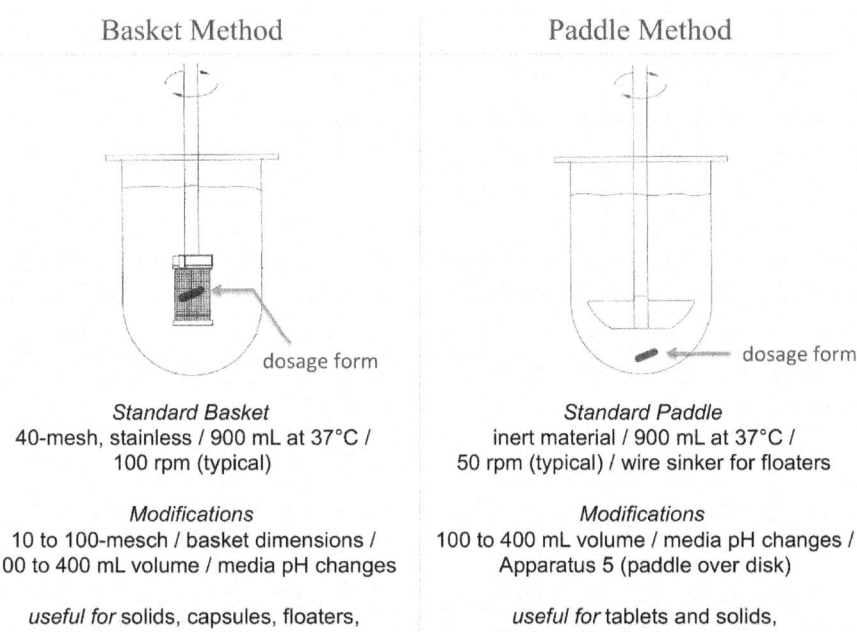

Figure 1.20: The schematic figure of the basket and paddle methods for dissolution testing (Hanson and Gray, 2004b).

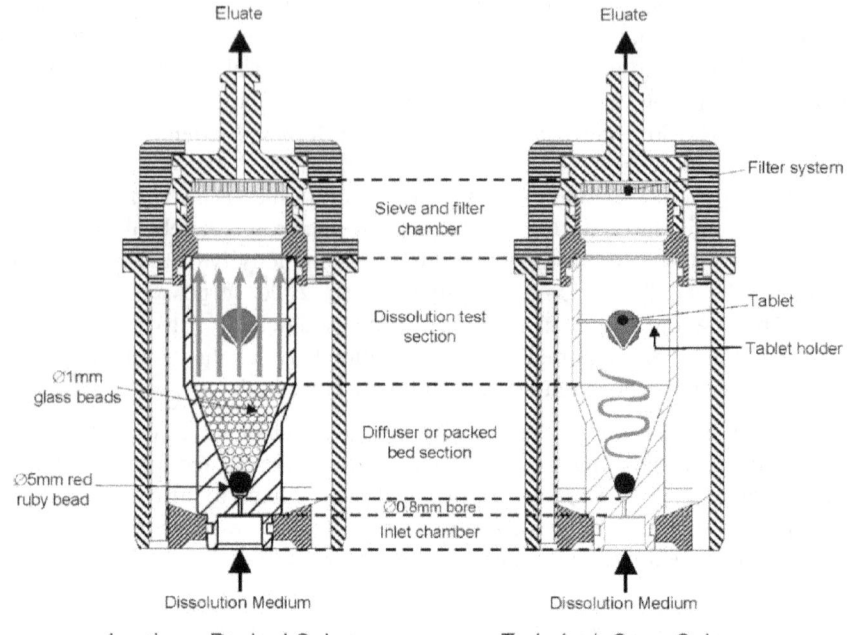

Figure 1.21: The schematic figure of the flow-through method, the flow-through cells in the open and packed mode, figure taken from Kakhi (2009a).

shear strain distribution that cannot be improved by increasing the rotation speed. The fluid mechanics in the flow-through cells are discussed extensively by Kakhi (2009a) and the fluid dynamics were successfully predicted by computational fluid dynamics (Kakhi, 2009b).

Dissolution Media

The dissolution measurements should be performed in dissolution media which can best predict the dissolution in vivo. Most common, buffer solutions are used to mimic gastric or intestinal pH. More sophisticated media include further ingredients like pepsin, lecithin, bile salts, or lipids. An update on the various dissolution media simulating proximal human gastrointestinal tract has been published by Jantratid et al. (2008). In case of poorly soluble drugs, sink condition can be obtained adding surfactants such as sodium dodecylsulfate (SDS) or sodium laurylsulfate (SLS) at concentrations above their critical micelle concentration. However, the opinions about the use of surfactants are divided. Tang et al. (2001) analyzed the bio-relevance of dissolution media containing SLS. Although a concentration of 1% resulted in sink conditions, the dissolution data obtained in vitro did not correlate with the in vivo data. Only lower concentrations of 0.25% SLS showed bio-relevant dissolution data, despite the over-saturated condition in the dissolution medium.

Evaluation of Dissolution Profiles (Costa and Sousa Lobo, 2001)

There are several mathematical models available to describe drug dissolution profiles. Depending on the focus, drug release can be analyzed as a function of the dosage form characteristics using mathematical models or by quantitative and statistical measures, the non-model dependent parameters. The mathematical models describing dissolution kinetics are often empirical equations. The common models are zero order, first order, Hixson-Crowell, Weibull, Higuchi, and Korsmeyer-Peppas. The common non-model dependent release parameters are dissolution time, assay time, dissolution efficiency, difference factor and similarity factor. These release parameters provide very limited information about the release mechanism and are suitable to compare dissolution profiles of same dosage form or to a reference.

The dissolution time $t_{x\%}$ presents the time necessary to release a determined percentage of drug, e.g., t_{15} or t_{60}. The assay time $t_{x\ min}$ presents the amount of drug dissolved at a specific time. Both parameters are very common in pharmacopoial monographs to determined the test limits, e.g., $t_{15} \geq 75\%$. Mean dissolution time and dissolution efficiency present further parameters to compare dissolution profiles. The mean dissolution time MDT is given by Equation (1.23),

$$MDT = \frac{\sum_{i=1}^{n} \hat{t}_i \cdot \Delta M_i}{\Delta M_i} \qquad (1.23)$$

where i is the sample number, n the number of dissolution sample times, $\hat{t}_i$ the time at midpoint between t_1 and t_{i-1}, and ΔM_i the additional amount of drug dissolved between t_i and t_{i-1}. The dissolution efficiency (DE) is described by the area under the dissolution curve AUC_{t_i} up to a certain time t_i expressed as a percentage of the area under the rectangle described by 100% dissolution in the same time c_{tot}.

$$DE = \frac{AUC_{t_i}}{c_{tot} t_i} \cdot 100 \qquad (1.24)$$

When comparing full dissolution profiles the multivariate analysis of variance (MAN-OVA) or the difference factor (f_1) and similarity factor (f_2) can be applied. f_1 and f_2 are calculated as of Equations (1.25) and (1.26),

$$f_1 = \left\{ \frac{\sum_{t=1}^{n} |R_t - T_t|}{\sum_{t=1}^{n} R_t} \right\} \cdot 100 \qquad (1.25)$$

$$f_2 = 50 \cdot \log \left\{ (1 + \frac{1}{n} \sum_{t=1}^{n} (R_t - T_t)^2 \right\}^{-0.5} \cdot 100) \qquad (1.26)$$

where n is the number of dissolution samplings, and R_t, T_t are the percent dissolved of the reference and test dissolution profile at each time point (Yuksel et al., 2000; FDA, 1997).

Bibliography

Aaltonen, J., Allesø, M., Mirza, S., Koradia, V., Gordon, K. C., Rantanen, J., 2009. Solid form screening – a review. Eur. J. Pharm. Biopharm. 71, 23–37.

Ahlneck, C., Zografi, G., 1990. The molecular basis of moisture effects on the physical and chemical stability of drugs in the solid state. Int. J. Pharm. 62, 87–95.

Alderborn, G., Nyström, C., 1996. Pharmaceutical powder compaction technology. Drugs and the pharmaceutical sciences. Marcel Dekker, Ch. Preface, pp. iii–v.

Alvarez-Lorenzo, C., Gómez-Amoza, J. L., Martínez-Pacheco, R., Souto, C., Concheiro, A., 2000. Evaluation of low-substituted hydroxypropylcelluloses (L-HPCs) as filler-binders for direct compression. Int. J. Pharm. 197, 107–116.

Ambrogi, V., Perioli, L., Marmottini, F., Accorsi, O., Pagano, C., Ricci, M., Rossi, C., 2008. Role of mesoporous silicates on carbamazepine dissolution rate enhancement. Micropor. Mesopor. Mat. 113, 445–452.

Amidon, G. L., Lennernäs, H., Shah, V. P., Crison, J. R., 1995. A theoretical basis for a biopharmaceutical drug classification: The correlation of in vitro drug product dissolution and in vivo bioavailability. Pharm. Res. 12, 413–420.

Armstrong, N. A., 2007a. Encyclopedia of Pharmaceutical Technology, 3rd Edition. Vol. 6. Informa Healthcare USA, Inc., Ch. Tablet Manufacture by Direct Compression, pp. 3673–3683.

Armstrong, N. A., 2007b. Encyclopedia of Pharmaceutical Technology, 3rd Edition. Vol. 6. Informa Healthcare USA, Inc., Ch. Tablet Manufacture, pp. 3653–3672.

Armstrong, N. A., 2008. Pharmaceutical dosage forms: Tablets. Rational design and formulation. Vol. 2 of Pharmaceutical Dosage Forms: Tablets. Informa Healthcare USA, Ch. Lubricants, Glidants, and Antiadherents, pp. 251–267.

Attia, M. A., Habib, F. S., 1985. Dissolution rates of carbamazepine and nitrazepam utilizing sugar solid dispersion system. Drug Dev. Ind. Pharm. 11, 1957–1969.

Augsburger, L. L., Hahm, H. A., Brzeczko, A. W., Shah, U., 2002. Encyclopedia of Pharmaceutical Technology. Informa Healthcare USA, Inc., Ch. Super Disintegrants: Characterization and Function, pp. 2623–2638.

Augsburger, L. L., Zellhofer, M. J., 2007. Encyclopedia of Pharmaceutical Technology, 3rd Edition. Vol. 6. Informa Healthcare USA, Inc., Ch. Tablet Formulation, pp. 3641–3652.

Behme, R. J., Brooke, D., 1991. Heat of fusion measurement of a low melting polymorph of carbamazepine that undergoes multiple-phase changes during differential scanning calorimetry analysis. J. Pharm. Sci. 80, 986–990.

Blagden, N., de Matas, M., Gavan, P. T., York, P., 2007. Crystal engineering of active pharmaceutical ingredients to improve solubility and dissolution rates. Adv. Drug Deliv. Rev. 59, 617–630.

Bolhuis, G. K., Armstrong, N. A., 2006. Excipients for direct compaction—An update. Pharm. Dev. Technol. 11, 111–124.

Bolourtchian, N., Nokhodchi, A., Dinarvand, R., 2001. The effect of solvent and crystallization conditions on habit modification of carbamazepine. Daru 9, 12–22.

Brittain, H., 1999. Polymorphism in pharmaceutical solids. Drugs and the pharmaceutical sciences. M. Dekker, Ch. The Phase Rules in Polymorphic Systems, pp. 35–72.

Brittain, H., Grant, D. J. W., 1999. Polymorphism in pharmaceutical solids. Drugs and the pharmaceutical sciences. M. Dekker, Ch. Effects of Polymorphism and Solid-state Solvation on Solubility and Dissolution Rate, pp. 279–330.

Brittain, H. G., 2004. Fluorescence studies of the transformation of carbamazepine anhydrate form lll to its dihydrate phase. J. Pharm. Sci. 93, 375–383.

Brown, M. B., 2005. The lost science of formulation. Drug Discovery Today 10, 1405–1407.

Burger, A., Henck, J.-O., Hetz, S., Rolling, J. M., Weissnicht, A. A., Stöttner, H., 2000. Energy/temperature diagram and compression behavior of the polymorphs of d-mannitol. J. Pharm. Sci. 89, 457–468.

Burger, A., Ramberger, R., 1979. On the polymorphism of pharmaceuticals and other molecular crystals. Microchimica Acta 72, 259–271.

Cammarn, S. R., Sakr, A., 2000. Predicting dissolution via hydrodynamics: salicylic acid tablets in flow through cell dissolution. Int. J. Pharm. 201, 199–209.

Carlin, B. A. C., 2008. Pharmaceutical dosage forms: Tablets. Rational design and formulation. Vol. 2 of Pharmaceutical Dosage Forms: Tablets. Informa Healthcare USA, Ch. Direct Compression and the Role of Filler-binder, pp. 173–216.

Carstensen, J. T., 2002. Modern Pharmaceutics. No. Bd. 121 in Drugs and the pharmaceutical sciences. Marcel Dekker, Ch. Preformulation, pp. 258–292.

Chieng, N., Rades, T., Aaltonen, J., 2011. An overview of recent studies on the analysis of pharmaceutical polymorphs. J. Pharm. Biomed. Anal. 55, 618–644.

Clas, S.-D., Dalton, C. R., Hancock, B. C., 1999. Differential scanning calorimetry: applications in drug development. Pharm. Sci. Technol. Today 2, 311–319.

Clas, S.-D., Dalton, C. R., Hancock, B. C., 2007. Encyclopedia of Pharmaceutical Technology, 3rd Edition. Vol. 1. Informa Healthcare USA, Inc., Ch. Calorimetry in Pharmaceutical Research and Development, pp. 393–405.

Costa, P., Sousa Lobo, J. M., 2001. Modeling and comparison of dissolution profiles. Eur. J. Pharm. Sci. 13, 123–133.

Cullity, B., Stock, S., 2001. Elements of X-ray diffraction. Pearson education. Prentice Hall, Ch. Geometry and Crystals, pp. 44–45.

Davidson, A., 1995. A multinational survey of the quatity of carbamazepine tablets. Drug Dev. Ind. Pharm. 21, 2167–2186.

Dürig, T., Fassihi, A., 1993. Identification of stabilizing and destabilizing effects of excipient-drug interactions in solid dosage form design. Int. J. Pharm. 97, 161–170.

Edwards, A. D., Shekunov, B. Y., Forbes, R. T., Grossmann, J. G., York, P., 2001. Time-resolved X-ray scattering using synchrotron radiation applied to the study of a polymorphic transition in carbamazepine. J. Pharm. Sci. 90, 1106–1114.

EL-Massik, M. A., Abdallah, O. Y., Galal, S., Daabis, N. A., 2006. Towards a universal dissolution medium for carbamazepine. Drug Dev. Ind. Pharm. 32, 893–905.

El-Zein, H., Riad, L., El-Bary, A. A., 1998. Enhancement of carbamazepine dissolution: in vitro and in vivo evaluation. Int. J. Pharm. 168, 209–220.

FDA, 1997. Guidance for industry: dissolution testing of immediate release solid oral dosage forms. Center for drug evaluation and research, Rockville, MD August.

FDA, September 2004. Guidance for industry: PAT — a framework for innovative pharmaceutical development, manufacturing, and quality assurance.

FDA, July 2007. Guidance for industry ANDAs: Pharmaceutical solid polymorphism chemistry, manufacturing, and controls information. U.S. Department of Health and Human Services Food and Drug Administration Center for Drug Evaluation and Research (CDER).

FDA CDER, 2009. The biopharmaceutics classification system (BCS) guidance. URL http://www.fda.gov/AboutFDA/CentersOffices/CDER/ucm128219.htm

Fleischman, S. G., Kuduva, S. S., McMahon, J. A., Moulton, B., Walsh, R. D. B., Rodrìguez-Hornedo, N., Zaworotko, M. J., 2003. Crystal engineering of the composition of pharmaceutical phases: multiple-component crystalline solids invloving carbamazepine. Crystal Growth and Des. 3, 909–919.

Florence, A. J., Johnston, A., Price, S. L., Nowell, H., Kennedy, A. R., Shankland, N., 2006. An automated parallel crystallisation search for predicted crystal structures and packing motifs of carbamazepine. J. Pharm. Sci. 95, 1918–1930.

Gift, A. D., Luner, P. E., Luedeman, L., Taylor, L. S., 2008. Influence of polymeric excipients on crystal hydrate formation kinetics in aqueous slurries. J. Pharm. Sci. 97, 5198–211.

Giron, D., 1995. Thermal analysis and calorimetric methods in the characterisation of polymorphs and solvates. Thermochim Acta 248, 1–59.

Gosselin, P., Thibert, R., Preda, M., McMullen, J., 2003. Polymorphic properties of micronized carbamazepine produced by ress. Int. J. Pharm. 252, 225–233.

Grant, D. J. W., 1999. Polymorphism in pharmaceutical solids. Drugs and the pharmaceutical sciences. M. Dekker, Ch. Theory and Origin of Polymorphism - Thermodynamics of Polymorphs, pp. 10–18.

Grzesiak, A. L., Lang, M., Kim, K., Matzger, A. J., 2003. Comparison of the four anhydrous polymorphs of carbamazepine and the crystal structure of form 1. J. Pharm. Sci. 92 (11), 2260–2271.

Han, J., Suryanarayanan, R., 1998. Influence of environmental conditions on the kinetics and mechanism of dehydration of carbamazepine dihydrate. Pharm. Dev. Technol. 3, 587–596.

Hanson, R., Gray, V., 2004a. Handbook of Dissolution Testing, 3rd Edition. Dissolution Technologies, Inc. Hockessin, Delaware, Ch. Theoretical Considerations, pp. 17–32.

Hanson, R., Gray, V., 2004b. Handbook of Dissolution Testing, 3rd Edition. Dissolution Technologies, Inc. Hockessin, Delaware, Ch. Dissolution Testing of Solid Dosage Forms, pp. 33–71.

He, X., 2008. Developing solid oral dosage forms: pharmaceutical theory and practice. Academic Press, Ch. Integration of Physical, Chemical, Mechanical, and Biopharmaceutical Properties in Solid Oral Dosage Form Development, pp. 409–442.

Hickey, M. B., Peterson, M. L., Scoppettuolo, L. A., Morrisette, S. L., Vetter, A., Guzmán, H., Remenar, J. F., Zhang, Z., Tawa, M. D., Haley, S., Zaworotko, M. J., Örn Almarsson, 2007. Performance comparison of a co-crystal of carbamazepine with marketed product. Eur. J. Pharm. Biopharm. 67, 112–119.

Hilfiker, R., 2006. Polymorphism in the pharmaceutical industry. Wiley-VCH, Ch. Preface, pp. xv–xvi.

Himes, V. L., Mighell, A. D., de Camp, W. H., 1981. Structure of carbamazepine: 5h-dibenz[b,f]azepine-5-carboxamide. Acta Cryst. B37, 2242–2245.

Jantratid, E., Janssen, N., Reppas, C., Dressman, J. B., 2008. Dissolution media simulating conditions in the proximal human gastrointestinal tract: An update. Pharm. Res. 25, 1663–1674.

Javadzadeh, Y., Jafari-Navimipour, B., Nokhodchi, A., 2007. Liquisolid technique for dissolution rate enhancement of a high dose water-insoluble drug (carbamazepine). Int. J. Pharm. 341, 26–34.

Jivraj, M., Martini, L. G., Thomson, C. M., 2000. An overview of the different excipients useful for the direct compression of tablets. Pharm. Sci. Technol. Today 3, 58–63.

Jørgensen, K., Jacobsen, L., 1992. Factorial design used for ruggedness testing of flow through cell dissolution method by means of weibull transformed drug release profiles. Int. J. Pharm. 88, 23–29.

Jung, H., Milán, R., Girard, M., León, F., Montoya, M., 1997. Bioequivalence study of carbamazepine tablets: in vitro/in vivo correlation. Int. J. Pharm. 152, 37–44.

Kakhi, M., 2009a. Classification of the flow regimes in the flow-through cell. Eur. J. Pharm. Sci. 37, 531–544.

Kakhi, M., 2009b. Mathematical modeling of the fluid dynamics in the flow-through cell. Int. J. Pharm. 376, 22–40.

Katzhendler, I., Azoury, R., Friedman, M., 1998. Crystalline properties of carbamazepine in sustained release hydrophilic matrix tablets based on hydroxypropyl methylcellulose. J. Control. Release 54, 69–85.

Katzhendler, I., Azoury, R., Friedman, M., 2000. The effect of egg albumin on the crystalline properties of carbamazepine in sustained release hydrophilic matrix tablets and in aqueous solutions. J. Control. Release 65, 331–343.

Kawashima, Y., Takeuchi, H., Hino, T., Niwa, T., Lin, T.-L., Sekigawa, F., Kawahara, K., 1993. Low-substituted hydroxypropylcellulose as a sustained-drug release matrix base or disintegrant depending on its particle size and loading in formulation. Pharmaceutical Research 10, 351–355.

Khankari, R. K., Grant, D. J., 1995. Pharmaceutical hydrates. Thermochim Acta 248, 61–79.

Kobayashi, Y., Ito, S., Itai, S., Yamamoto, K., 2000. Physicochemical properties and bioavailability of carbamazepine polymorphs and dihydrate. Int. J. Pharm. 193, 137–146.

Kogermann, K., Aaltonen, J., Strachan, C. J., Heinämäki, K. P. J., Yliruusi, J., Rantanen, J., 2008. Establishing quantitative in-line analysis of multiple solid-state transformations during dehydration. J. Pharm. Sci. 97, 4983–4998.

Koivisto, M., Heinänen, P., Tanninen, V. P., Lehto, V.-P., 2006. Depth profiling of compression-induced disorders and polymorphic transition on tablet surfaces with grazing incidence X-ray diffraction. Pharm. Res. 23, 813–820.

Kompendium, 2011. Arzneimittel-Kompendium der Schweiz.
URL http://www.kompendium.ch/

Kukura, J., Baxter, J., Muzzio, F., 2004. Shear distribution and variability in the USP Apparatus 2 under turbulent conditions. Int. J. Pharm. 279, 9–17.

Laine, E., Tuominen, V., Ilvessalo, P., Kahela, P., 1984. Formation of dihydrate from carbamazepine anhydrate in aqueous conditions. Int. J. Pharm. 20, 307–314.

Lake, O. A., Olling, M., Barends, D. M., 1999. In vitro/in vivo correlations of dissolution data of carbamazepine immediate release tablets with pharmacokinetic data obtained in healthy volunteers. Eur. J. Pharm. Biopharm. 48 (1), 13–19.

Lang, M., Kampf, J. W., Matzger, A. J., 2002. Form IV of Carbamazepine. J. Pharm. Sci. 91, 1186–1190.

Lefebvre, C., Guyot-Hermann, A. M., Draguet-Brughmans, M., Bouché, R., Guyot, J. C., 1986. Polymorphic transitions of carbamazepine during grinding and compression. Drug Dev. Ind. Pharm. 12, 1913–1927.

Li, Y., Chow, P. S., Tan, R. B. H., Black, S. N., 2008. Effect of water activity on the transformation between hydrate and anhydrate of carbamazepine. Org. Process Res. Dev. 12, 264–270.

Lindenberg, M., Kopp, S., Dressman, J. B., 2004. Classification of orally administered drugs on the world health organization model list of essential medicines according to the biopharmaceutics classification system. Eur. J. Pharm. Biopharm. 58, 265–278.

Luhtala, S., 1992. Effect of sodium lauryl sulphate and polysorbate 80 on crystal growth and aqueous solubility of carbamazepine. Acta Pharm. Nordica 4, 1100–1801.

Mauger, J., Ballard, J., Brockson, R., De, S., Gray, V., Robinson, D., August 2003. Intrinsic dissolution performance testing of the USP dissolution Apparatus 2 (rotating paddle) using modified salicylic acid calibrator tablets: Proof of principle. Dissolution Technologies 10 (3).

Meyer, M. C., Staughn, A. B., Mhatre, R. M., Shah, V. P., Williams, R. L., Lesko, L. J., 1998. The relative bioavailability and in vivo - in vitro correlations for four marketed carbamazepine tablets. Pharm. Res. 15 (11), 1787–1791.

Meyer, M. C., Straughn, A. B., Jarvi, E. J., Wood, G. C., Pelsor, F. R., Shah, V. P., 1992. The bioinequivalence of carbamazepine tablets with a history of clinical failures. Pharm. Res. 9, 1612–1616.

Mittapalli, P. K., Suresh, B., Hussaini, S. S. Q., Rao, Y. M., Apte, S., 2008. Comparative in vitro study of six carbamazepine products. AAPS PharmSciTech 9 (2), 357.

Moneghini, M., Kikic, I., Voinovich, D., Perissutti, B., Filipović-Grčić, J., 2001. Processing of carbamazepine–PEG 4000 solid dispersions with supercritical carbon dioxide: preparation, characterisation, and in vitro dissolution. Int. J. Pharm. 222, 129–138.

Moreton, R. C., 2008. Pharmaceutical dosage forms: Tablets. Rational design and formulation. Vol. 2 of Pharmaceutical Dosage Forms: Tablets. Informa Healthcare USA, Ch. Disintegrants in Tablets, pp. 217–249.

Nair, R., Gonen, S., Hoag, S. W., 2002. Influence of polyethylene glycol and povidone on the polymorphic transformation and solubility of carbamazepine. Int. J. Pharm. 240, 11–22.

Nokhodchi, A., Bolourtchian, N., Dinarvand, R., 2005. Dissolution and mechanical behaviors of recrystallized carbamazepine from alcohol solution in the presence of additives. J. Cryst. Growth 274, 573–584.

O'Brien, L. E., Timmins, P., Williams, A. C., York, P., 2004. Use of in situ FT-Raman spectroscopy to study the kinetics of the transformation of carbamazepine polymorphs. J. Pharm. Biomed. Anal. 36, 335–340.

Ono, M., Tozuka, Y., Oguchi, T., Yamamura, S., Yamamoto, K., 2002. Effects of dehydration temperature on water vapor adsorption and dissolution behavior of carbamazepine. Int. J. Pharm. 239, 1–12.

Otsuka, M., Ishii, M., Matsuda, Y., 2003. Effect of surface modification on hydration kinetics of carbamazepine anhydrate using isothermal microcalorimetry. AAPS PharmSciTech 4, 1–9.

Otsuka, M., Ohfusa, T., Matsuda, Y., 2000. Effect of binders on polymorphic transformation kinetics of carbamazepine in aqueous solution. Colloid Surface 17, 145–152.

Papadopoulou, V., Valsami, G., Dokoumetzidis, A., Macheras, P., 2008. Biopharmaceutics classification systems for new molecular entities (BCS-NMEs) and marketed drugs (BCS-MD): Theoretical basis and practical examples. Int. J. Pharm. 361, 70–77.

Patterson, J. D., Bailey, B. C., 2007. Solid-State Physics: Introduction to the Theory, 2nd Edition. Springer, Ch. 1.2.5 List of Crystal Systems and Bravais Lattices (B), pp. 23–25.

Peltonena, L., Liljerothb, P., Heikkilä, T., Kontturib, K., Hirvonen, J., 2003. Dissolution testing of acetylsalicylic acid by a channel flow method — correlation to USP basket and intrinsic dissolution methods. Eur. J. Pharm. Sci. 19, 395–401.

Picker-Freyer, K. M., Dürig, T., 2007. Physical mechanical and tablet formation properties of hydroxypropylcellulose: In pure form and in mixtures. AAPS PharmSciTech 8 (4).

Qu, H., Louhi-Kultanen, M., Kallas, J., 2006. Solubility and stability of anhydrate/hydrate in solvent mixtures. Int. J. Pharm. 321, 101–107.

Qureshi, S. A., 2007. Encyclopedia of Pharmaceutical Technology, 3rd Edition. Vol. 6. Informa Healthcare USA, Inc., Ch. Tablet Testing, pp. 3707–3716.

Roberts, R., Rowe, R., 1996. Influence of polymorphism on the young's modulus and yield stress of carbmazepine, sulfathiazole and sulfanilamide. Int. J. Pharm. 129, 79–94.

Roberts, R. J., Payne, R. S., Rowe, R. C., 2000. Mechanical property predictions for polymorphs of sulphathiazole and carbamazepine. Eur. J. Pharm. Sci. 9, 277–283.

Rockland, L. B., Beuchat, L. R. (Eds.), 1986. Water activity: theory and applications in food: [proceedings of the tenth basic symposium held in Dallas]/[sponsored by the Institute of Food Technologists (IFT), Chicago, and the International Union of Food Science and Technology]. New York; Basel: M. Dekker, cop. 1987.

Rodrìguez-Hornedo, N., Kelly, R. C., Sinclair, B. D., Miller, J. M., 2007. Encyclopedia of Pharmaceutical Technology, 3rd Edition. Vol. 2. Informa Healthcare USA, Inc., Ch. Crystallization: General Principles and Significance on Product Development, pp. 834–857.

Rodrìguez-Hornedo, N., Murphy, D., 2004. Surfactant-facilitated crystallization of dihydrate carbamazepine during dissolution of anhydrous polymorph. J. Pharm. Sci. 93, 449–460.

Rodrìguez-Sponga, B., Priceb, C. P., Jayasankara, A., Matzger, A. J., Rodrìguez-Hornedo, N., 2004. General principles of pharmaceutical solid polymorphism: a supramolecular perspective. Adv. Drug Deliv. Rev. 56, 241–274.

Rowe, R. C., Sheskey, P. J., , Owen, S. C. (Eds.), 2006. Handbook of Pharmaceutical Excipients, Part 3, 5th Edition. Pharmaceutical Press.

Rustichelli, C., Gamberini, G., Ferioli, V., Gamberini, M., Ficarra, R., Tommasini, S., 2000. Solid-state study of polymorphic drugs: carbamazepine. J. Pharm. Biomed. Anal. 23, 41–54.

Salameh, A. K., Taylor, L. S., 2006. Physical stability of crystal hydrates and their anhydrates in the presence of excipients. J. Pharm. Sci. 95, 446–461.

Schindler, W., Häfliger, F., 1954. Über Derivate des Iminodibenzyls. Helvetica Chimica Acta 37, 472–483.

Schmidt, P. C., 2007. Encyclopedia of Pharmaceutical Technology, 3rd Edition. Vol. 5. Informa Healthcare USA, Inc., Ch. Secondary Electron Microscopy in Pharmaceutical Technology, pp. 3217–3256.

Schulz, M., Fussnegger, B., Bodmeier, R., 2010. Adsorption of carbamazepine onto crospovidone to prevent drug recrystallization. Int. J. Pharm. 391, 169–176.

Sethia, S., Squillante, E., 2004. Solid dispersion of carbamazepine in PVP K30 by conventional solvent evaporation and supercritical methods. Int. J. Pharm. 272, 1–10.

Singhal, D., Curatolo, W., 2004. Drug polymorphism and dosage form design: a practical perspective. Adv. Drug Deliv. Rev. 56, 335–347.

Stahly, G. P., 2007. Diversity in single- and multiple-component crystals. The search for and prevalence of polymorphs and cocrystals. J. Cryst. Growth 7, 1007–1026.

Summers, M. P., Enever, B. P., Carless, J. E., 1977. Influence of crystal form on tensile strength of compacts of pharmaceutical materials. Journal of Pharmaceutical Sciences 66, 1172–1175.

Suryanarayanan, R., Rastogi, S., 2007. Encyclopedia of Pharmaceutical Technology, 3rd Edition. Vol. 6. Informa Healthcare USA, Inc., Ch. X-Ray Powder Diffractometry, pp. 4103–4116.

Tang, L., Khan, S. U., Muhammad, N. A., 2001. Evaluation and selection of biorelevant dissolution media for a poorly water- soluble new chemical entity. Pharm. Dev. Technol. 6, 531–540.

Tian, F., Sandler, N., Gordon, K. C., McGoverin, C. M., Reay, A., Strachan, C. J., Saville, D. J., Rades, T., 2006. Visualizing the conversion of carbamazepine in aqueous suspension with and without the presence of excipients: A single crystal study using SEM and Raman microscopy. Eur. J. Pharm. Biopharm. 64, 326–335.

Tian, F., Zhang, F., Sandler, N., Gordon, K., McGoverin, C., Strachan, C., Saville, D., Rades, T., 2007. Influence of sample characteristics on quantification of carbamazepine hydrate formation by X-ray powder diffraction and Raman spectroscopy. Eur. J. Pharm. Biopharm. 66, 466–474.

Tsinman, K., Avdeef, A., Tsinman, O., Voloboy, D., 2009. Powder dissolution method for estimating rotating disk intrinsic dissolution rates of low solubility drugs. Pharm. Res. 26, 2093–2100.

Ugaonkar, S., Nunes, A. C., Needham, T. E., 2007. Effect of n-scCO2 on crystalline to amorphous conversion of carbamazepine. Int. J. Pharm. 333, 152–161.

USP 31, 2008. USP 31; United States Pharmacopoeia / National Formulary. United States Pharmacopeial Convention.

Vippagunta, S. R., Brittain, H. G., Grant, D. J., 2001. Crystalline solids. Adv. Drug Deliv. Rev. 48, 3–26.

Šehić, S., Betz, G., Hadžidedić, Š., El-Arini, S. K., Leuenberger, H., 2010. Investigation of intrinsic dissolution behavior of different carbamazepine samples. Int. J. Pharm. 386, 77–90.

Wang, J., Flanagan, D. R., 2008. Developing solid oral dosage forms: pharmaceutical theory and practice. Academic Press, Ch. Fundamentals of Dissolution, pp. 309–318.

Wang, J., Shiu, G., Ting, O., Viswanathan, C., Skelly, J., 1993. Effects of humidity and temperature on in-vitro dissolution of carbamazepine tablets. J. Pharm. Sci. 82, 1002–1005.

Wells, J., Aulton, M. E., 2007. Aulton's pharmaceutics: the design and manufacture of medicines, 3rd Edition. Churchill Livingstone, Ch. Pharmaceutical preformulation, pp. 336–360.

Wesselingh, J. A., Frijlink, H. W., 2008. Pharmaceutical dosage forms: Tablets. Rational design and formulation. Pharmaceutical Dosage Forms: Tablets. Informa Healthcare USA, Ch. Mass Transfer from Solid Oral Dosage Forms, pp. 22–36.

Wrolstad, R. E., Decker, E. A., Schwartz, S. J., Sporns, P., 2005. Handbook of Food Analytical Chemistry, Water, Proteins, Enzymes, Lipids, and Carbohydrates. Wiley-IEEE, Ch. A2.3 Measurement of Water Activity Using Isopiestic Method, p. 51.

York, P., 1983. Solid-state properties of powders in the formulation and processing of solid dosage forms. Int. J. Pharm. 14, 1–28.

Yu, L., 2001. Amorphous pharmaceutical solids: preparation, characterization and stabilization. Adv. Drug Deliv. Rev. 48, 27–42.

Yu, L. X., Carlin, A. S., Amidon, G. L., Hussain, A. S., 2004. Feasibility studies of utilizing disk intrinsic dissolution rate to classify drugs. Int. J. Pharm. 270, 221–227.

Yuksel, N., KanÄśk, A. E., Baykara, T., 2000. Comparison of in vitro dissolution profiles by ANOVA-based, model-dependent and -independent methods. Int. J. Pharm. 209, 57–67.

Zakeri-Milani, P., Barzegar-Jalali, M., Azimi, M., Valizadeha, H., 2009. Biopharmaceutical classification of drugs using intrinsic dissolution rate (IDR) and rat intestinal permeability. Eur. J. Pharm. Biopharm. 73, 102–106.

Zeitler, J. A., Taday, P. F., Gordon, K. C., Pepper, M., Rades, T., 2007. Solid-state transition mechanism in carbamazepine polymorphs by time-resolved terahertz spectroscopy. ChemPhysChem 8, 1924–1927.

Zerrouk, N., Toscani, S., Gines-Dorado, J.-M., Chemtob, C., Ceólin, R., Dugué, J., 2001. Interactions between carbamazepine and polyethylene glycol (PEG) 6000: characterisations of the physical, solid dispersed and eutectic mixtures. Eur. J. Pharm. Sci. 12, 395–404.

Zhang, G. G., Law, D., Schmitt, E. A., Qiu, Y., 2004. Phase transformation considerations during process development and manufacture of solid oral dosage forms. Adv. Drug Deliv. Rev. 56, 371–390.

Chapter 2

Objectives

Physicochemical properties of raw materials can vary among drug suppliers with consequence on the drug performance in the final dosage form. Depending on the way of production, the raw materials show impurities in form of solvent inclusions or polymorphic mixtures, and particle morphology can be remarkably different among the various raw materials.

Carbamazepine (CBZ) is one example where variability in raw materials has been linked to irregular dissolution behavior, bioinequivalence, and clinical failures of the final dosage form. The purpose of this study was to investigate the impact of variability in CBZ raw materials on the drug release. Hence, the aims of the study include the following issues:

- Polymorphic and morphological characterization of CBZ raw materials of four different suppliers and investigation on intrinsic dissolution behavior of the samples.
- Evaluation of the effect of two commonly used tablet fillers on the variability of CBZ raw materials.
- Investigation on the recrystallization of CBZ samples to reduce variability.
- Development of a tablet formulation that is robust towards the variability in CBZ samples and that conforms to the USP requirements of CBZ tablets for immediate release.
- Proposition of a strategy to control the variability in CBZ raw materials.

Chapter 3

Original Publications

3.1 Variability in Commercial Carbamazepine Samples – Impact on Drug Release

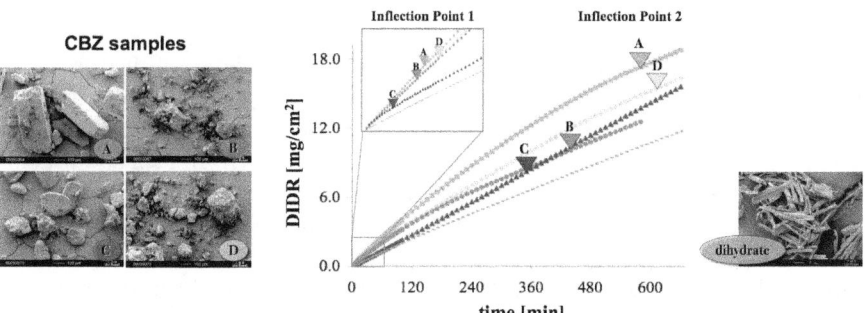

Abstract

The aim of this study was to characterize the variability of commercial carbamazepine (CBZ) samples and to investigate the influence of two commonly used tablet fillers, i.e., mannitol and microcrystalline cellulose (MCC) on the CBZ sample variability. Polymorphism and morphology of CBZ samples were analyzed by differential scanning calorimetry, X-ray powder diffraction, sieve analysis, and scanning electron microscopy. Drug release from CBZ samples and binary mixtures (30–90% drug load) was characterized by a unidirectional dissolution method measuring disk intrinsic dissolution rate (DIDR) and drug release, respectively.

All CBZ samples were of p-monoclinic form but differed in their polymorphic purity, particle size, morphology, and intrinsic dissolution rate. Two characteristic inflection points, determined in the DIDR profiles, characterized the specific transformation behavior of each CBZ sample. The variability in CBZ samples was also exhibited in the drug release profiles from their binary mixtures. Mannitol increased initial drug release of CBZ samples up to 10-fold in mixtures of 30% drug load. The presence of MCC resulted in reduced variability in drug release. The unidirectional dissolution method is presented as a straightforward monitoring tool to characterize variability of CBZ raw materials and effect of tablet fillers.

Keywords: Carbamazepine, Dihydrate, Preformulation, Morphology, Polymorphism, Tablet filler.

Introduction

Carbamazepine (CBZ) is a well-established drug to treat epilepsy and trigeminal neuralgia and there are already several generics on the market. Nevertheless, CBZ products present a history of irregular drug performance and clinical failures. Several reports show high dissolution variability in CBZ tablets on the market world-wide and even among CBZ tablets of the same brand (Meyer et al., 1992, 1998; Davidson, 1995; Al-Zein et al., 1999; Lake et al., 1999; Mittapalli et al., 2008).

CBZ is poorly soluble in water with a narrow therapeutic index, classified as class II drug according to the Biopharmaceutics Classification System (Lindenberg et al., 2004). Furthermore, CBZ exhibits at least four polymorphic forms referring to p-monoclinic (form III), triclinic (form I), c-monoclinic (form IV), and trigonal (form II) crystal lattice. At room temperature CBZ form III is the thermodynamically most stable form and also the form commercially available. CBZ further exists as several solvates, e.g., CBZ monoacetonate (Fleischman et al., 2003). Both, CBZ polymorphs and solvates show differences in crystal form and consequently also difference in melting point, solubility, compactability, and chemical reactivity.

Information about the processes used in raw drug production is generally limited. Polymorphic form may be the same among various commercial sources, however, different solvents and additives can be used for the crystallization of drug raw materials. These dissimilar processes lead to different crystal habits and possible solvent inclusions, which result in changed solubility and dissolution behavior of CBZ crystals (Mahalaxmi et al., 2009; Blagden et al., 2007; Bolourtchian et al., 2001). Furthermore, physical processes, such as milling or grinding, can cause crystal defects, mechanical activation, and small amount of amorphous, which then also lead to faster and inhomogeneous dissolution behavior of CBZ samples (Tian et al., 2006b; Murphy et al., 2002; Mosharraf et al., 1999; Lefebvre et al., 1986).

In aqueous media, anhydrous CBZ transforms into its dihydrate. This transformation critically influences dissolution and bioavailability of CBZ formulations and it has been in the focus of many investigations over the last 20 years. The mechanism of transformation is a solution-mediated process where dihydrate is formed by whisker growth, clearly visible by optical microscopy (Laine et al., 1984). Raman spectroscopy and fluorescence studies are presented as non-invasive methods to study the transformation behavior of CBZ (Tian et al., 2006a,b; Brittain, 2004). Furthermore, the influence of transformation on the dissolution rate has been investigated. A decrease in dissolution rate is expected as soon as the less soluble CBZ dihydrate is formed. This phenomenon is detected as an inflection point in the disk intrinsic dissolution rate (DIDR) profile of CBZ samples (Kobayashi et al., 2000; Šehić et al., 2010). Kobayashi et al. (2000) reported that the transformation depends on the polymorphic form. CBZ form I transforms faster to dihydrate than form III in vitro and results in lower bioavailability in vivo. Šehić et al. (2010), however, reported different transformation points within commercial CBZ samples, though same polymorphic form was specified. This difference also shows in tablet formulation using Ludipress® as tablet filler. To date, DIDR profiles focused on the onset of transformation only.

Composition of the dissolution media also influences the transformation of CBZ. Dihydrate formation is strongly inhibited in solutions (1–4%, w/v) of hydroxypropyl methylcellulose (HPMC), hydroxypropylcellulose (HPC), or polyvinylpyrrolidone (PVP) whereas solutions of polyvinyl alcohol (PVA), polyethylene glycol (PEG), and sodium carboxymethylcellulose (CMC) have only weak inhibitory activity (Gift et al., 2008; Tian et al., 2006a; Otsuka et al., 2000). In contrast, surfactants, such as sodium lauryl sulphate and sodium taurocholate promote the dihydrate formation during the dissolution test (Rodrìguez-Hornedo and Murphy, 2004). Nevertheless, the knowledge about the effect of commonly used tablet fillers is still limited (Salameh and Taylor, 2006).

In this study, a unidirectional dissolution method was developed building on the DIDR method of Šehić et al. (2010) to characterize CBZ samples by DIDR profiles of the transformation range and to investigate their initial drug release of binary mixtures with commonly used tablet fillers. For this purpose, two different types of direct compacted filler were selected, mannitol as water soluble and microcrystalline cellulose (MCC) as water insoluble tablet filler.

Materials and Methods

CBZ samples were obtained from four different suppliers; for confidential reasons they were designated as CBZ A, B, C, and D. All samples were of commercial grade, no further specification for the drug substance was made from our side. Samples were stored at room temperature (20-25 °C) and controlled relative humidity (43% RH). *CBZ dihydrate* was prepared according to McMahon et al. (1996). Anhydrous CBZ was stirred in water for 24 h. Crystals were filtered under suction and dried at ambient conditions over night. The obtained CBZ dihydrate was kept at room temperature and 80% RH.

Excipients for the binary mixtures: Mannitol (Parteck® M300, MERCK KGaA, Germany) and microcrystalline cellulose (MCC SANAQ® 102L, Pharmatrans SANAQ AG, Switzerland) were used as the tablet fillers. All other chemicals and reagents purchased from commercial sources were of analytical grade.

Morphological Characterization

For morphological characterization scanning electron microscopy (SEM) (ESEM XL 30 FEG, Philips, The Netherlands) was applied at a voltage of 10 kV and magnifications of 100–2000 times. Before analysis, powder was sprinkled on carbon adhesive, and compacts were fixed to the sample holder by conductive silver. The samples were then sputtered with gold.

Sieve analysis was performed with 100 g samples on standard sieving tower (Retsch Type Vibro, Schieritz & Hauenstein AG, Switzerland) and 10 min shaking time at 40 Hz.

Polymorphic Characterization

Polymorphic form of CBZ samples was characterized by X-ray powder diffractometry (XRPD) using a diffractometer (D5000, Siemens, Germany). The powder was filled into special holders and the surface was pressed flat. Operating conditions were Ni filtered Cu Kα radiation (λ =1.5406), 40 kV, and 30 mA. Step was 0.02° 2θ, step time 1.0 s, angular scanning speed 1° 2θ/min, and angular range between 5° and 40° 2θ scale.

Thermal behavior of all samples was analyzed by differential scanning calorimetry (DSC), using heat flux DSC (4000, PerkinElmer, USA). DSC was calibrated with indium prior to the measurement. A sample of 3–6 mg was accurately weighted into an aluminum pan with holes and scanned between 40 °C and 220 °C at 10 °C/min under dry nitrogen gas purge (20 ml/min).

True density of all samples was assessed by a gas displacement pycnometer (AccuPyc 1330, Micromeritics, USA). Powder was purged with helium by five repetitive purging cycles and the density was reported as average value. The test was performed in triplicate.

Water Activity Measurement

Water activity (a_W) is the partial vapor pressure (p) of a sample relative to the vapor pressure of pure water (p_0). It is equal to the equilibrium relative humidity (ERH) of air over the sample, e.g., powder or tablets (Wrolstad et al., 2005); see Equation (3.1). All parameters are temperature dependent.

$$a_W = \frac{p}{p_0} = ERH \tag{3.1}$$

Water activity of all samples was monitored with the digital water activity analyzer (Hygropalm, Rotronic AG, Switzerland) at $23.5 \pm 1.5\,°C$.

Intrinsic Dissolution of CBZ Samples

CBZ samples were analyzed by disk intrinsic dissolution. Intrinsic dissolution rate (IDR) is the specific dissolution rate of a pure drug from one surface only. IDR is described by Equation (3.2),

$$IDR = \frac{dC}{dt} \times \frac{V}{S} = k \times C_S \tag{3.2}$$

where $\frac{dC}{dt}$ is the amount of drug dissolving over time, S the surface area of the compact, k the intrinsic dissolution constant, and C_S the solubility of the drug. Disk intrinsic dissolution curves are linear and the slope is the amount of drug dissolved per cm^2. Under fixed experimental conditions IDR value is characteristic for a given substance. It is used in preformulation studies to classify and predict possible problems in bioavailability of a drug (Zakeri-Milani et al., 2009; Hanson and Gray, 2004).

Sample Preparation. Flat-faced compacts of 400 mg and 0.95 cm^2 surface area were produced from CBZ samples and CBZ dihydrate using a material tester (Zwick 1478, Germany). Compact porosity was controlled to $11.3 \pm 0.7\%$ and hardness was 20–41 N. For compacts of CBZ A, porosity was 16%, as higher compaction forces resulted in lamination. Compaction force was set between 6 and10 kN, while compaction, decompaction, and ejection speed were set at 10 mm/min, 50 mm/min, and 10 mm/min, respectively.

Porosity of CBZ compacts was measured by a mercury porosimeter (AutoPore IV 9500, Micrometrics, USA). Pressure in the range of 0.014–0.227 MPa corresponding to the mean pore diameter of 6 nm to 400 μm, was applied. Surface tension (γ) of mercury was 485 dynes/cm and contact angle (θ) was 130°.

Unidirectional Dissolution Method. To obtain disk intrinsic dissolution rate (DIDR) profiles of the CBZ samples unidirectional dissolution was performed by a modified USP Apparatus I. The compact was placed in a sample holder fitting to the rotating unit of the dissolution apparatus (SotaxAT7*smart*, SOTAX AG, Switzerland) and the compact was embedded in melted paraffin that only one surface was available to the dissolution media. Unidirectional dissolution method was performed at two conditions: one for the analysis of first inflection point referring to the start of transformation, and another for the second inflection

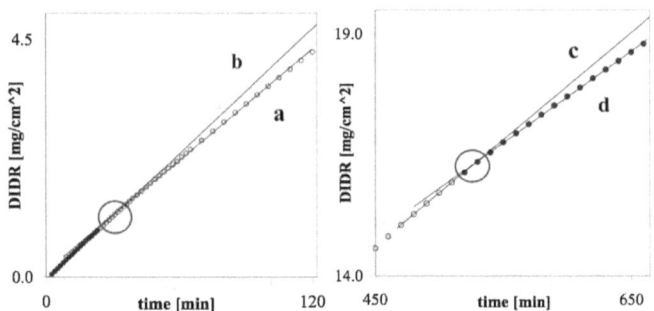

Figure 3.1: Inflection point in DIDR profiles of 2-h run (left) and 10–11-h run (right) determined by the intercept of slope a and b, and slope c and d, respectively. In the 10–11-h run only every second data point is shown.

point referring to the stabilized transformation rate. Release media was 500 ml and 1000 ml water, and run time was 2 h and 10–11 h for these two conditions, respectively. Dissolution media was at 37 ± 0.5 °C and a rotation speed of 100 rpm was selected as it showed the lowest standard deviation for drug release of each sample. Drug content of the media was measured at predetermined time intervals by UV-VIS spectrophotometer (Lambda 25, PerkinElmer, USA) at 285 nm.

Evaluation of DIDR Profiles. IDR was determined by the slope in the DIDR profile prior to the first inflection point, the slope referring to the release of anhydrous CBZ before transforming to the dihydrate.

The inflection point was determined by the intercept of two linear regressions with the best fit ($R^2 \geq 0.999$). The linear regressions a and b in the DIDR profiles of 2-h run and c and d in the profiles of 10-11-h run were used to determine the first and second inflection point, respectively (Figure 3.1). For the first inflection point, approximation of the best linear regression (a) included the initial DIDR profile starting from 2.6 min DIDR measurement. The linear regression b was set through all data points of the later stage in the DIDR profile that were not included in the regression line a. For the second inflection point, approximation of the best linear regression (d) included the late stage in DIDR profile, where release rate stabilized. The linear regression c was set through the ten preceding release values not included in regression d.

Binary Mixtures

CBZ samples were mixed with mannitol or MCC for 10 min in a mixer (Turbula® type T2C, W. Bachofen, Switzerland) and for a further 2 min after adding 1% magnesium stearate (Sandoz, Switzerland) to the mixture. Drug loads were set at 90%, 70%, 50%, and 30% and mixtures of 20 g were prepared. Binary mixtures were compacted by the same procedure as the compaction of CBZ samples at the condition of 50 mm/min, 100 mm/min, and 50 mm/min for compaction, decompaction, and ejection speed, respectively. Porosity was kept at approximately 12% based on tablet volume. To compare drug release of the binary mixtures and CBZ samples, initial drug release of binary mixtures was tested by the dissolution method of 2 h (Section Unidirectional Dissolution Method). Drug release profiles of binary mixtures were compared by the amount of drug dissolved after 120 min dissolution test. For all statistical comparison, one-way ANOVA followed by Student's t-test was applied.

Results and Discussion

Morphological Characterization

With SEM a difference in morphology and surface texture was clearly visible among the CBZ samples (Figure 3.2). CBZ A showed very crystalline particles of prismatic shape, whereas CBZ B and D showed particles with rough and uneven edges and a lot of fine particles. Samples of CBZ C were more round and with smoother surface. Prismatic shape was reported characteristic for CBZ form III (Krahn and Mielck, 1987). CBZ dihydrate powders were of fine needle-like structure in loose aggregates as expected (Šehić et al., 2010).

Table 3.1: Sieve fractions of CBZ in [%] total sample weight (n = 3).

[μm]	CBZ A	CBZ B	CBZ C	CBZ D
≥710	0.5	0.1	3.1	2.8
500–710	3.5	0.1	11.8	26.5
355–500	10.7	1.2	11.3	17.2
250–355	23.8	16.2	15.4	17.5
180–250	20.9	23.0	10.4	10.3
125–180	21.3	24.9	20.0	7.9
90–125	12.1	13.2	15.3	6.9
≤90	7.3	21.4	12.8	11.0
Average	233 ± 5	166 ± 4	241 ± 21	333 ± 32
Median	209 ± 9	155 ± 4	192 ± 8	331 ± 20

CBZ samples also differed in particle size distribution, in average and in median particle size (Table 3.1). CBZ A and D showed mono-modal particle size distribution with a very broad peak found in CBZ D. For CBZ B and C, particle size distribution was bi-modal showing two maxima at 0–90 μm and 125–180 μm,

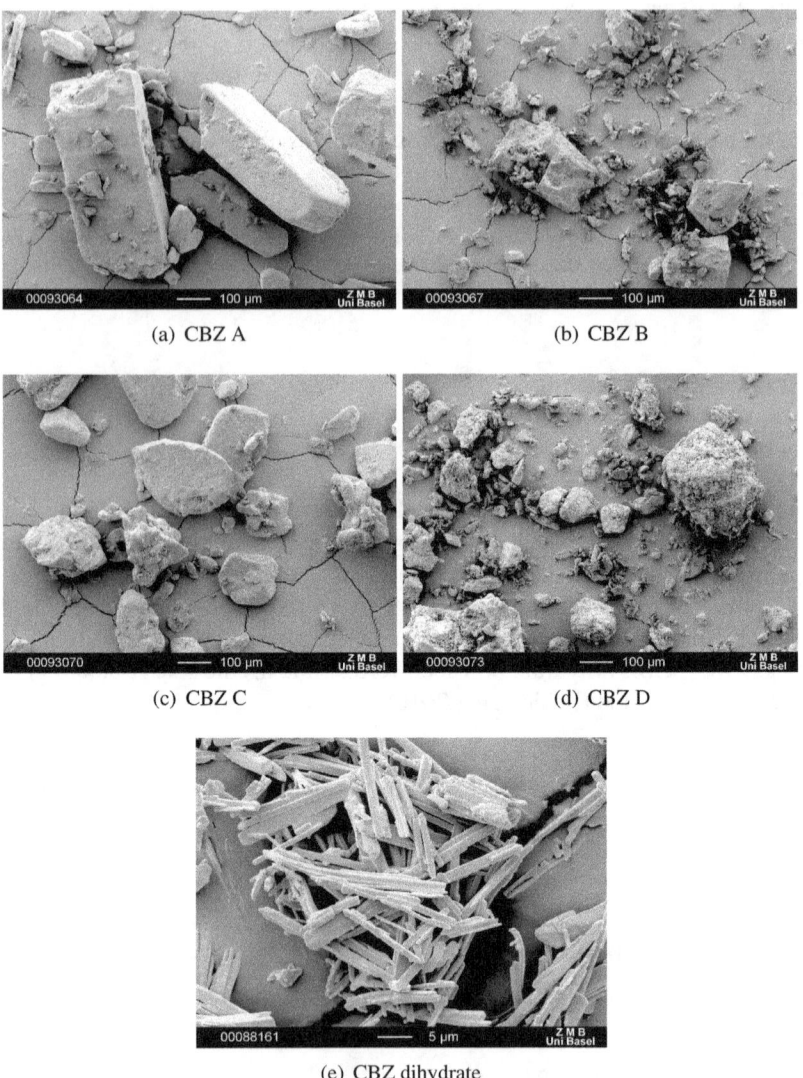

Figure 3.2: SEM pictures of CBZ samples and CBZ dihydrate.

and at 125-180 μm and 250–355 μm, respectively. All CBZ samples except for CBZ A contained fines (particle size $\leq$ 90 μm) of at least 11% with the ranking of CBZ B > C > D> A.

Polymorphic Characterization

XRPD results are shown in Figure 3.3. The diffractograms of all CBZ samples showed the characteristic peaks reported for CBZ p-monoclinic (Getsoian et al., 2008; Grzesiak et al., 2003). Peaks at angles smaller than 10°, indicating presence of CBZ trinclinic or the trigonal form, were not detected. However, there were some indications as to the presence of form IV. There are peaks at 12.68° and 29.91° 2θ with increasing intensity for CBZ B, C, and D. Diffractograms of CBZ D further showed an extra peak at 19.15°, and at 22.91° 2θ (see arrows). The presence of CBZ form IV in commercial CBZ samples was suspected previously (Kipouros et al., 2005). At lower angles of 5–20° 2θ, XRPD results showed difference in peak intensities, which could be due to different degree in crystallinity, smaller particle size, or due to preferred orientation of the particles in the sample holder. XRPD data did not show any change in the diffraction pattern in CBZ samples after compaction (data not shown). CBZ dihydrate was also confirmed by XRPD. Characteristic peaks for CBZ dihydrate were at 2θ = 8.9°, 18.9°, and 19.4°, and they are in agreement with the literature (Kobayashi et al., 2000).

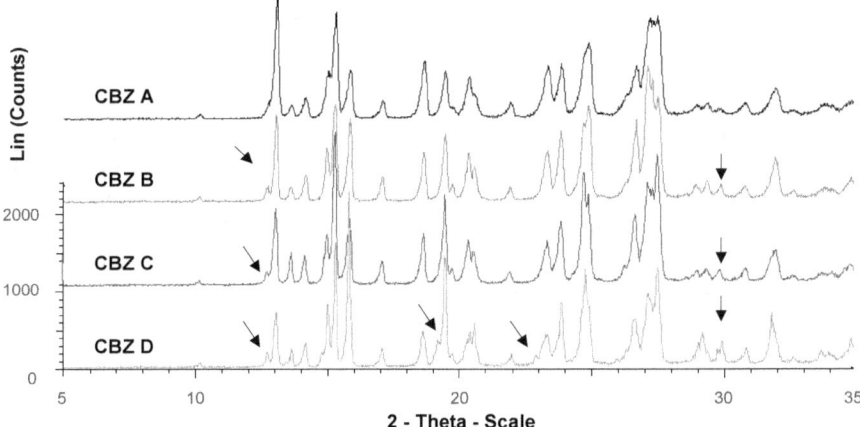

Figure 3.3: X-ray powder diffraction of CBZ samples; 5–40° 2θ at 1° 2θ/min.

DSC results are presented in Figure 3.4. Most CBZ samples showed the characteristic thermal events of CBZ form III as reported previously (Grzesiak et al., 2003). The onset of each thermal event (T_{onset}) and the enthalpy (ΔH) were used to compare the endothermic and exothermic peaks among the CBZ samples. Melting of form III was visible around 176 °C, followed by crystallization to form I around 182 °C and melting of form I around 191 °C. However, DSC profiles of CBZ samples differed visibly. Thermograms of CBZ A and B showed the characteristic peaks for form III clearly. However, there was a difference in melting enthalpy around 176 °C (form III), whereas melting enthalpy around 191 °C (form I) was the same. This indicates presence of form I in CBZ B. Nonetheless, only small amounts

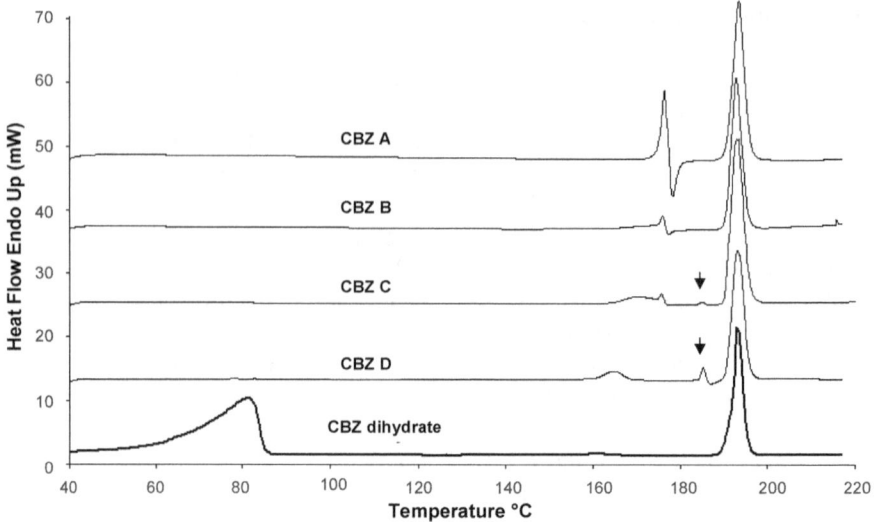

Figure 3.4: DSC profiles of CBZ samples and CBZ dihydrate; 40–220 °C at 10 °C/min

can be assumed, as no specific peaks for form I were detected by XRPD. Further difference among CBZ samples was seen for CBZ D, where the endothermic peak around 176 °C was missing and an endotherm showed at lower temperature of around 164 °C instead. For both, CBZ C and D samples a further endothermic peak was visible at 185 °C (arrows) indicating presence of CBZ form IV (Grzesiak et al., 2003). This finding is consistent with the XRPD results, where small amounts of form IV were suspected in both CBZ C and D. However, for CBZ B, no indications of form IV were detected by DSC. It has to be noted here, that the detection limit for CBZ polymorphic form is reported to be the best with 1% and 5–10% for Hyper-DSC and XRPD, respectively (McGregor et al., 2004; Lefebvre et al., 1986). Furthermore, thermograms for CBZ B and C showed a shoulder to the endothermic peak around 176 °C starting around 164 °C. This shoulder has been identified to be due to a direct solid–solid conversion of Form III to form I (Zeitler et al., 2007). Deviations in thermal behavior can further be due to solvent inclusion in crystal pores, contamination by other polymorphic form, and amorphous traces. CBZ dihydrate had two endothermic events in DSC measurements. Dehydration was visible as a wide endothermic peak in temperature range of 50–85 °C. Second endothermic peak was at 192 °C showing melting of CBZ from I.

True density of all CBZ samples was 1.338 ± 0.001 g/cm^3 indicating CBZ polymorphic form III (Krahn and Mielck, 1989; Grzesiak et al., 2003).

Water Activity

For all CBZ anhydrous samples water activity was between 0.241 and 0.423. At a specific water activity hydration of drug can occur (Qiu et al., 2009). Also the phase conversion of anhydrous CBZ to CBZ dihydrate depends on the water activity of the system. In the water activity range of 0.601–0.641 at 20 °C, CBZ form III and CBZ dihydrate are reported to coexist in equilibrium (Qu et al., 2006; Li et al., 2008). Our samples were therefore below the critical range for transformation.

DIDR Profiles of CBZ Samples

DIDR profiles of the four CBZ samples and the CBZ dihydrate are shown in Figure 3.5 and Table 3.2 presents the IDR values. Initial phase in DIDR profiles showed a significant difference in IDR for each CBZ samples ($p < 0.05$). IDR was in the rank of CBZ A>D>B>C and the IDR values were in the same range (29, 35.5, and 37.6 $\mu g/cm^2/min$) as reported previously (Yu et al., 2004; Zakeri-Milani et al., 2009; Murphy et al., 2002). IDR value of CBZ dihydrate was 1.2–2.2 times lower than for anhydrous CBZ samples. The comparatively low dissolution rate of CBZ C might be explained by presence of small amounts of the less soluble CBZ form IV and compact hardness. CBZ C compacts showed breaking strength about 2 times higher than of the other compacts.

Table 3.2: Intrinsic dissolution rate of CBZ samples and CBZ dihydrate (CBZ dihyr) in the initial DIDR values prior to the first inflection point.

Sample $n \geq 3$	IDR [$\mu g/cm^2/min$]
CBZ A	43.9 ± 3.1
CBZ B	38.6 ± 0.8
CBZ C	24.0 ± 0.4
CBZ D	39.9 ± 0.7
CBZ dihyr	19.5 ± 0.7

It is interesting to note, that the amount of drug dissolved after 120 min inversely correlated with the amount of fines in the CBZ samples. Generally, samples with higher amount of fines show faster dissolution rate, in case of CBZ, however, the opposite was the case. This leads to the conclusion, that CBZ fines transformed faster to the slower dissolving CBZ dihydrate and therefore decreased the actual amount of drug dissolved after 120 min.

Inflection Point. DIDR profiles showed not only one point of transformation at the early stage of dissolution, but transformation continued over a time range

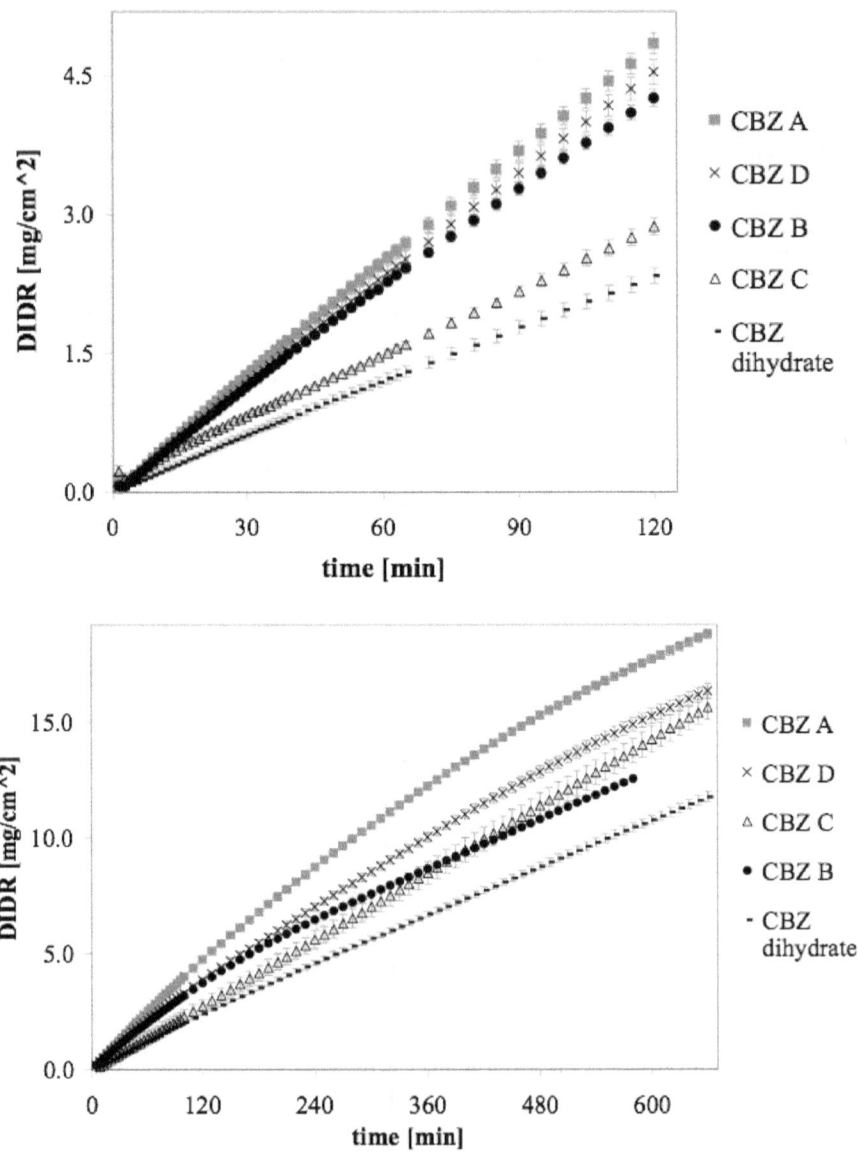

Figure 3.5: DIDR profiles of pure carbamazepine from suppliers A, B, C, and D; 2 h (top) and 10–11 h (bottom) dissolution time. Average values (n≥3) are presented.

Table 3.3: First and second inflection point in DIDR profiles of CBZ samples.

Sample n ≥ 3	First inflection point [min]	Second inflection point [min]
CBZ A	33 ± 6	570 ± 25
CBZ B	28 ± 10	443 ± 93
CBZ C	16 ± 3	350 ± 11
CBZ D	41 ± 7	635 ± 19

stabilizing at a later stage within 10 h. The results for the first and second inflection points are shown in Table 6.5. First inflection points were in the same range as reported previously (Šehić et al., 2010; Kobayashi et al., 2000). CBZ samples differed significantly in their first and second inflection point ($p < 0.05$). Both inflection points could be ranked as C≤B<A<D. The ranking was significant only for the second inflection point ($p < 0.05$), except for CBZ B over C ($p = 0.059$). CBZ B showed the highest standard deviation for both inflection points, which might be due to the high amount of fines.

The two inflection points could describe the transformation process from CBZ anhydrous to dihydrate in dissolution. The first inflection point reflected the onset of the gradual change in dissolution rate induced by the forming and dissolution of CBZ dihydrate. The second inflection point reflected onset of the constant dissolution rate, where the transformation process and dissolution were in equilibrium. Transformation as a process over time has been studied on CBZ compact surface during intrinsic dissolution by in situ Raman spectroscopy (Lehto et al., 2008). Based on the knowledge of previous studies gained by Raman spectroscopy, fluorescence studies, or XRPD (Tian et al., 2006a; Brittain, 2004; Suryanarayanan, 1989), the transformation range could be derived from the DIDR profiles and thereby characterize the CBZ samples. To confirm the formation of CBZ dihydrate compact surfaces were analyzed by SEM at several time points in DIDR test as described in the following section.

Surface Change of CBZ Compacts During DIDR Test. To avoid recrystallization due to sample preparation, the CBZ compacts were removed from the samples holder inclusive the paraffin layer and gently placed upside down on a kitchen paper for a few seconds. Samples were then kept under controlled RH of 43% for a maximum of 24 h hours prior the the analysis. SEM pictures of the compact surfaces after 30, 60, 120, and 480 min DIDR test confirmed CBZ dihydrate formation (Figure 3.6). The needle-shaped dihydrate crystals on compact surface were visible at a later stage in DIDR test than their onset of dissolution in the DIDR profiles. Dihydrate was first visible on the CBZ B and D compacts after 60 min and on CBZ A and C compacts after 120 min. CBZ dihydrate crystals grow at disrupted crystal structure of CBZ anhydrate induced, e.g., by grinding (Murphy et al., 2002). CBZ B and D showed the most irregular particle shapes in SEM, they were therefore expected to have the most crystal defects and transform the fastest. For CBZ D, however, the first inflection point showed later compared to the other samples. A reason could be that CBZ D showed less fines compared to CBZ B. The amount of fines may be indicative for crystal defects in a sample, as fines can be caused by milling or grinding.

Binary Mixtures

DSC results for binary mixtures are shown in Figure 3.7. An effect of tablet filler on the thermal behavior of CBZ was observed. In presence of MCC, the intensity of the thermal events was reduced and broken peaks were visible around 190 °C. In presence of mannitol, a change in the thermal behavior of CBZ was observed. Mannitol melted around 164 °C and CBZ seemed to dissolve fully in melted mannitol of 30% drug load and partially at the higher drug loads resulting in a broadening of melting point for CBZ form III and a shift towards lower temperature around 172 °C compared to melting of pure form III at 176 °C. The same mixtures did not show any change in crystallinity in XRPD results (data shown in Appendix 6.1.8). Therefore, the interaction with mannitol is limited to temperatures of 164 °C and above. Also Joshi et al. (2002) reported the interaction of CBZ with mannitol as a temperature related phenomenon detected in DSC.

Drug Release of Binary Mixtures. Drug release profiles of the binary mixtures comparing CBZ samples at same drug load are presented in Figure 3.8 and 3.9 shows the comparison of drug release of different drug loads with CBZ B as an example. An overview of amount of CBZ released from the binary mixtures after 120 min dissolution test is shown in Table 3.4.

It is interesting to note that the variability in DIDR profiles found in the CBZ samples was present also in the drug release profiles of the binary mixtures. Amount of CBZ released after 120 min differed significantly among CBZ samples of the same drug load (ANOVA; $p < 0.001$, except for the 50% CBZ–MCC mixture). Difference in amount of drug dissolved among the various CBZ samples was much less with the presence of MCC than of mannitol. Methods for the analysis of dissolution curves were found to be either over discriminative, i.e., analysis by one-way repeated measures ANOVA (MANOVA), or not discriminative enough, i.e., difference factor (f_1) and similarity factor (f_2) proposed by FDA (1997).

Drug release profiles of binary mixtures showed dependence on drug load. Dissolution rate was increased at the lower drug loads of 50% and 30%, especially in binary mixtures with mannitol. Amount of drug dissolved was 2 times higher for compacts of 50% and 7–10 times higher for compacts of 30% drug load with mannitol. For all binary mixtures with 90% drug load, drug release was slower than for pure CBZ compacts.

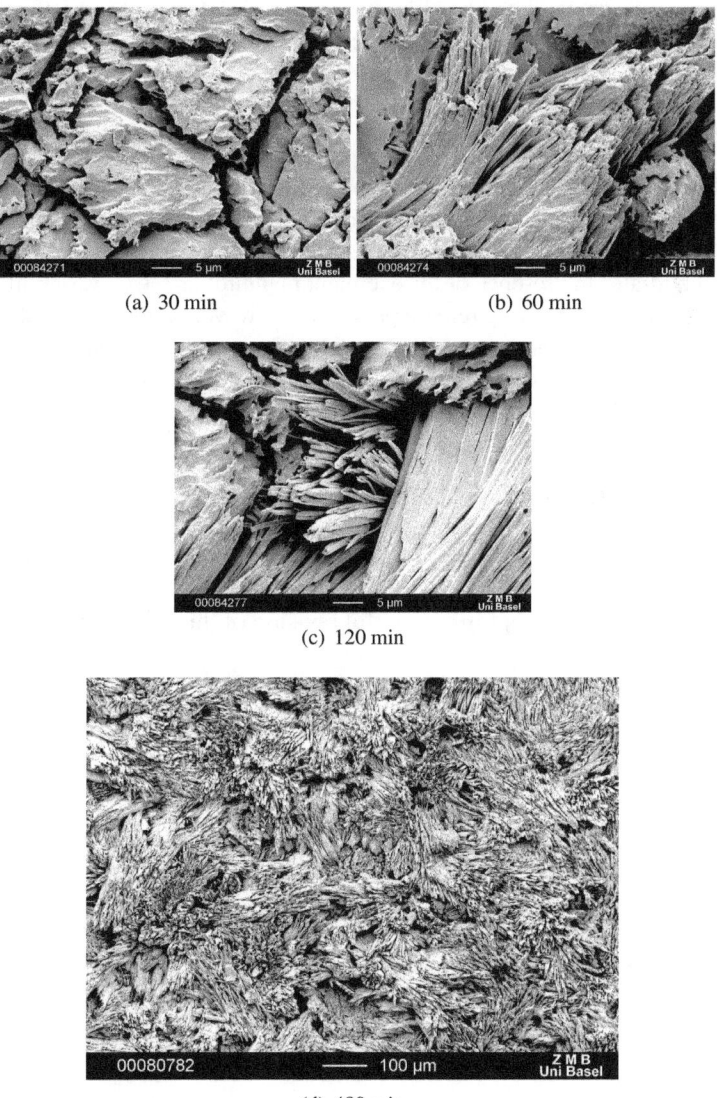

(a) 30 min (b) 60 min

(c) 120 min

(d) 480 min

Figure 3.6: SEM pictures of compact surface from CBZ B during DIDR test.

A possible reason for the different effects of mannitol and MCC on drug release could be their solubility difference. Mannitol dissolved very fast compared to the drug. CBZ samples could express their dissolution variability, even at the lower drug loads, where mannitol enhanced the solubility of CBZ. In contrast, compacts with the insoluble MCC showed swollen surfaces with wide cracks after the drug release measurements. The fiber-like structure of MCC could form a matrix that was able to control the amount of CBZ released.

Conclusions

The unidirectional dissolution method proved to be a straightforward monitoring tool for preformulation studies. Commercial CBZ samples could be specified by two inflection points in the DIDR profiles reflecting the individual transformation to CBZ dihydrate. In presence of the excipient mannitol or MCC, variability among CBZ samples regarding drug release persisted. However, high amount of mannitol strongly increased release rate of CBZ, and MCC was able to reduce the variability of CBZ samples. To control drug release and bioavailability of CBZ in tablet formulation, MCC is suggested as a suitable tablet filler.

Acknowledgments

The financial support by the Senglet Stiftung Basel, Switzerland, in form of a Ph.D. scholarship for Felicia Flicker is kindly appreciated. Acknowledgment has to be given to Dr. K. Chansanroj for her careful revision of this manuscript.

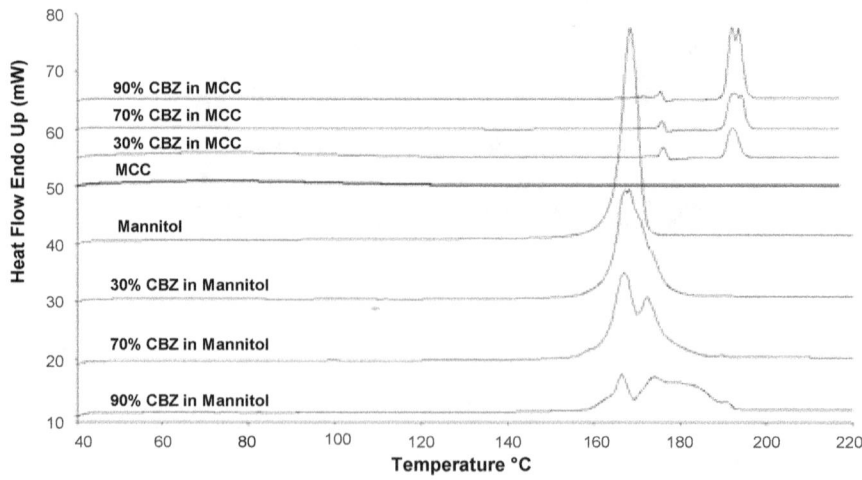

Figure 3.7: DSC profiles of binary mixtures of CBZ B in MCC and mannitol; 40 °C–220 °C at 10 °C/min

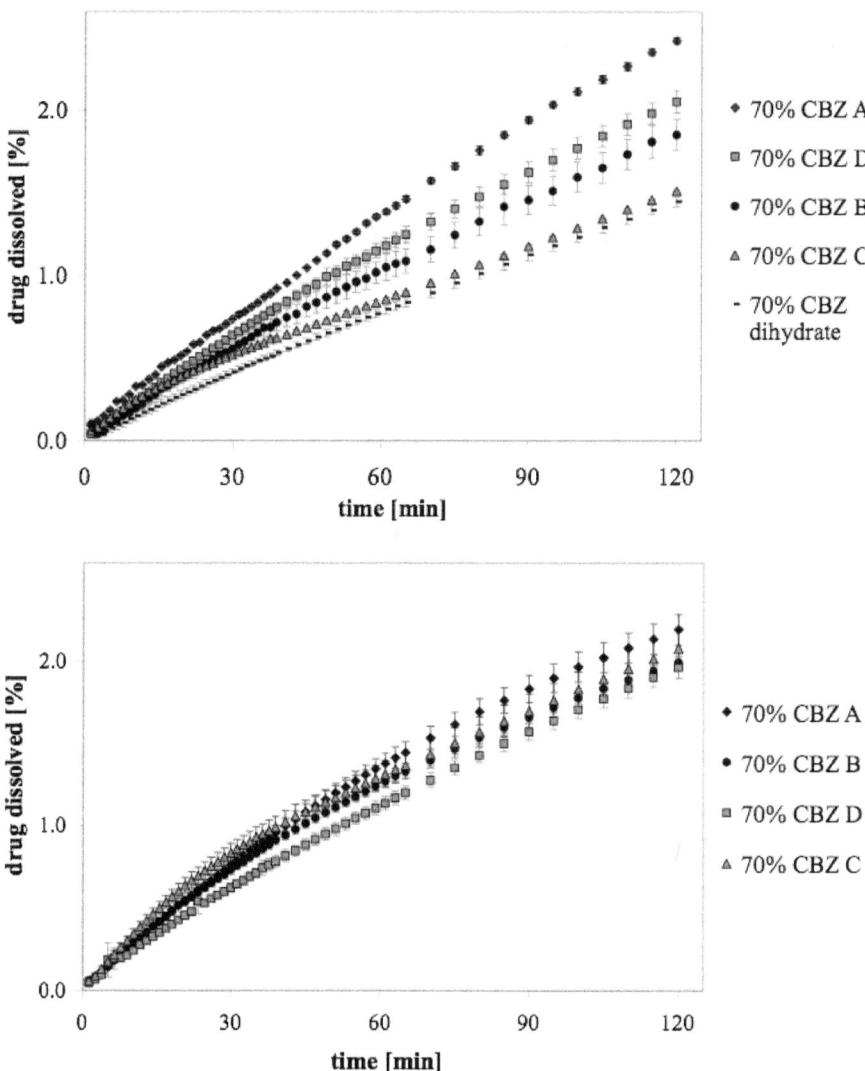

Figure 3.8: Initial drug release profiles of binary mixtures at 70% drug load with mannitol (top) and with MCC (bottom).

Table 3.4: Amount of CBZ dissolved [%] from binary mixtures with mannitol and MCC at different drug loads after 120 min.

	Mannitol	MCC
30% CBZ		
A	16.06 ± 2.95	4.39 ± 0.22
B	10.73 ± 0.73	3.67 ± 0.27
C	13.39 ± 1.71	4.98 ± 0.08
D	10.61 ± 1.51	4.41 ± 0.28
50% CBZ		
A	3.70 ± 0.18	2.49 ± 0.56
B	3.28 ± 0.24	3.30 ± 0.37
C	2.58 ± 0.18	2.92 ± 0.59
D	3.33 ± 0.10	3.31 ± 0.39
70% CBZ		
A	1.85 ± 0.09	2.19 ± 0.09
B	2.05 ± 0.07	1.99 ± 0.04
C	2.43 ± 0.02	2.08 ± 0.12
D	1.52 ± 0.01	1.96 ± 0.07
90% CBZ		
A	2.00 ± 0.09	3.23 ± 0.26
B	1.70 ± 0.05	1.47 ± 0.09
C	1.47 ± 0.03	1.22 ± 0.01
D	1.63 ± 0.06	1.58 ± 0.10
100% CBZ		
A	2.33 ± 0.03	
B	2.18 ± 0.02	
C	1.39 ± 0.02	
D	2.17 ± 0.03	

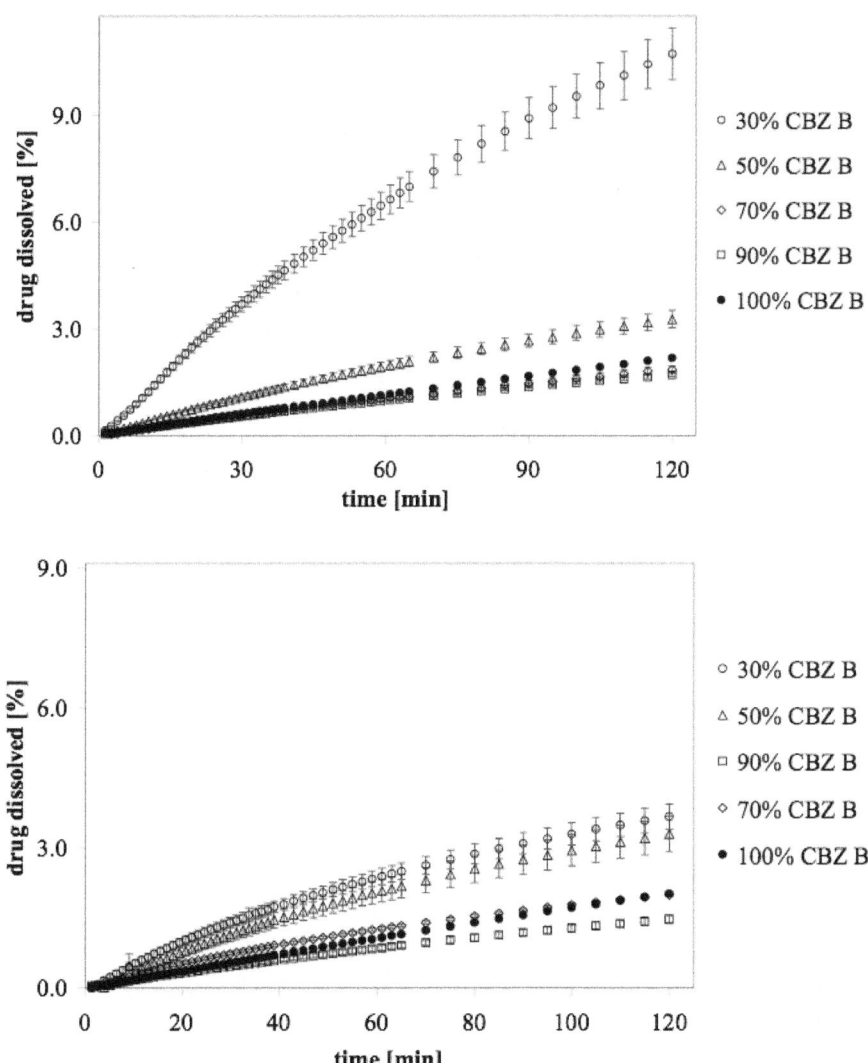

Figure 3.9: Initial drug release profiles of binary mixture with mannitol (top) and with MCC (bottom).

Bibliography

Al-Zein, H., Riad, L. E., Abd-Elbary, A., 1999. Effect of packaging and storage on the stability of carbamazepine tablets. Drug Dev. Ind. Pharm. 25 (2), 223–227.

Blagden, N., de Matas, M., Gavan, P. T., York, P., 2007. Crystal engineering of active pharmaceutical ingredients to improve solubility and dissolution rates. Adv. Drug Deliv. Rev. 59, 617–630.

Bolourtchian, N., Nokhodchi, A., Dinarvand, R., 2001. The effect of solvent and crystallization conditions on habit modification of carbamazepine. Daru 9, 12–22.

Brittain, H. G., 2004. Fluorescence studies of the transformation of carbamazepine anhydrate form lll to its dihydrate phase. J. Pharm. Sci. 93, 375–383.

Davidson, A., 1995. A multinational survey of the quatity of carbamazepine tablets. Drug Dev. Ind. Pharm. 21 (19), 2167–2186.

FDA, 1997. Guidance for industry: dissolution testing of immediate release solid oral dosage forms. Center for drug evaluation and research, Rockville, MD August.

Fleischman, S. G., Kuduva, S. S., McMahon, J. A., Moulton, B., Walsh, R. D. B., Rodrìguez-Hornedo, N., Zaworotko, M. J., 2003. Crystal engineering of the composition of pharmaceutical phases: multiple-component crystalline solids invloving carbamazepine. Crystal Growth and Design 3 (6), 909–919.

Getsoian, A., Lodaya, R., Blackburn, A., 2008. One-solvent polymorph screen of carbamazepine. Int. J. Pharm. 348, 3–9.

Gift, A. D., Luner, P. E., Luedeman, L., Taylor, L. S., 2008. Influence of polymeric excipients on crystal hydrate formation kinetics in aqueous slurries. J. Pharm. Sci. 97 (12), 5198–211.

Grzesiak, A. L., Lang, M., Kim, K., Matzger, A. J., 2003. Comparison of the four anhydrous polymorphs of carbamazepine and the crystal structure of form 1. J. Pharm. Sci. 92 (11), 2260–2271.

Hanson, R., Gray, V., 2004. Handbook of Dissolution Testing, 3rd Edition. Dissolution Technologies, Inc. Hockessin, Delaware, Ch. Dissolution testing of solid dosage forms, pp. 33–71.

Joshi, B. V., Patil, V. B., Pokharkar, V. B., 2002. Compatibility studies between carbamazepine and tablet excipients using thermal and non-thermal methods. Drug Dev. Ind. Pharm. 28 (6), 687–694.

Kipouros, K., Kachrimanis, K., Nikolakakis, I., Malamataris, S., 2005. Quantitative analysis of less soluble form iv in commercial carbamazepine (form iii) by diffuse reflectance fourier transform spectroscopy (drifts) and lazy learning algorithm. Anal Chim Acta 550, 191–198.

Kobayashi, Y., Ito, S., Itai, S., Yamamoto, K., 2000. Physicochemical properties and bioavailability of carbamazepine polymorphs and dihydrate. Int. J. Pharm. 193, 137–146.

Krahn, F. U., Mielck, J. B., 1987. Relation between several polymorphic forms and the dihydrate of carbamazepine. Pharm. Acta Helv 62, 247–254.

Krahn, F. U., Mielck, J. B., 1989. Effect of type and extent of crystalline order on chemical and physical stability of carbamazepine. Int. J. Pharm. 53, 25–34.

Laine, E., Tuominen, V., Ilvessalo, P., Kahela, P., 1984. Formation of dihydrate from carbamazepine anhydrate in aqueous conditions. Int. J. Pharm. 20, 307–314.

Lake, O. A., Olling, M., Barends, D. M., 1999. In vitro/in vivo correlations of dissolution data of carbamazepine immediate release tablets with pharmacokinetic data obtained in healthy volunteers. Eur. J. Pharm. Biopharm. 48 (1), 13–19.

Lefebvre, C., Guyot-Hermann, A. M., Draguet-Brughmans, M., Bouch, R., Guyot, J. C., 1986. Polymorphic transitions of carbamazepine during grinding and compression. Drug Dev. Ind. Pharm. 12, 1913–1927.

Lehto, P., Aaltonen, J., Tenho, M., Rantanen, J., Hirvonen, J., Tanninen, V. P., Peltonen, L., 2008. Solvent-mediated solid phase transformations of carbamazepine: effects of simulated intestinal fluid and fasted state simulated intestinal fluid. J. Pharm. Sci. 98 (3), 985–996.

Li, Y., Chow, P. S., Tan, R. B. H., Black, S. N., 2008. Effect of water activity on the transformation between hydrate and anhydrate of carbamazepine. Org. Process Res. Dev. 12, 264–270.

Lindenberg, M., Kopp, S., Dressman, J. B., 2004. Classification of orally administered drugs on the world health organization model list of essential medicines according to the biopharmaceutics classification system. Eur. J. Pharm. Biopharm. 58, 265–278.

Mahalaxmi, R., Ravikumar, Pandey, S., Shirwaikar, A., Shirwaikar, A., 2009. Effect of recrystallization on size, shape, polymorph and dissolution of carbamazepine. Int. J. PharmTech Res. 1 (3), 725–732.

McGregor, C., Saunders, M. H., Buckton, G., Saklatvala, R. D., 2004. The use of high-speed differential scanning calorimetry (hyper-dsc) to study the thermal properties of carbamazepine polymorphs. Thermochim Acta 417, 231–237.

McMahon, L. E., Timmins, P., Williams, A. C., York, P., 1996. Characterization of dihydrates prepared from carbamazepine polymorphs. J. Pharm. Sci. 85 (10), 1064–1069.

Meyer, M. C., Staughn, A. B., Mhatre, R. M., Shah, V. P., Williams, R. L., Lesko, L. J., 1998. The relative bioavailability and in vivo - in vitro correlations for four marketed carbamazepine tablets. Pharm. Res. 15 (11), 1787–1791.

Meyer, M. C., Straughn, A. B., Jarvi, E. J., Wood, G. C., Pelsor, F. R., Shah, V. P., 1992. The bioinequivalence of carbamazepine tablets with a history of clinical failures. Pharm. Res. 9, 1612–1616.

Mittapalli, P. K., Suresh, B., Hussaini, S. S. Q., Rao, Y. M., Apte, S., 2008. Comparative in vitro study of six carbamazepine products. AAPS PharmSciTech 9 (2), 357.

Mosharraf, M., Sebhatu, T., Nyström, C., 1999. The effects of disordered structure on the solubility and dissolution rates of some hydrophilic, sparingly soluble drugs. Int. J. Pharm. 177, 29–51.

Murphy, D., Rodríguez-Cintrón, F., Langevin, B., Kelly, R. C., Rodríguez-Hornedo, N., 2002. Solution-mediated phase transformation of anhydrous to dihydrate carbamazepine and the effect of lattice disorder. Int. J. Pharm. 246, 121–134.

Otsuka, M., Ohfusa, T., Matsuda, Y., 2000. Effect of binders on polymorphic transformation kinetics of carbamazepine in aqueous solution. Colloid Surface 17, 145–152.

Qiu, Y., Chen, Y., Liu, L., Zhang, G. G. Z. (Eds.), 2009. Developing solid oral dosage forms: pharmaceutical theory and practice. Academic Press, Ch. 2.3.2 Solvates/Hydrates, pp. 32–34.

Qu, H., Louhi-Kultanen, M., Kallas, J., 2006. Solubility and stability of anhydrate/hydrate in solvent mixtures. Int. J. Pharm. 321, 101–107.

Rodrìguez-Hornedo, N., Murphy, D., 2004. Surfactant-facilitated crystallization of dihydrate carbamazepine during dissolution of anhydrous polymorph. J. Pharm. Sci. 93, 449–460.

Salameh, A. K., Taylor, L. S., 2006. Physical stability of crystal hydrates and their anhydrates in the presence of excipients. J. Pharm. Sci. 95, 446–461.

Suryanarayanan, R., 1989. Determination of the relative amounts of anhydrous carbamazepine ($C_{15}H_{12}N_2O$) and carbamazepine dihydrate ($C_{15}H_{12}N_2O\ 2H_2O$) in a mixture by powder X-ray diffractometry. Pharmaceutical Research 6, 1017–1024.

Tian, F., Sandler, N., Gordon, K. C., McGoverin, C. M., Reay, A., Strachan, C. J., Saville, D. J., Rades, T., 2006a. Visualizing the conversion of carbamazepine in aqueous suspension with and without the presence of excipients: A single crystal study using SEM and Raman microscopy. Eur. J. Pharm. Biopharm. 64, 326–335.

Tian, F., Zeitler, J., Strachan, C., Saville, D., Gordon, K., Rades, T., 2006b. Characterizing the conversion kinetics of carbamazepine polymorphs to the dihydrate in aqueous suspension using Raman spectroscopy. J. Pharm. Biomed. Anal. 40, 271–280.

Šehić, S., Betz, G., Hadžidedić, Š., El-Arini, S. K., Leuenberger, H., 2010. Investigation of intrinsic dissolution behavior of different carbamazepine samples. Int. J. Pharm. 386, 77–90.

Wrolstad, R. E., Decker, E. A., Schwartz, S. J., Sporns, P., 2005. Handbook of Food Analytical Chemistry, Water, Proteins, Enzymes, Lipids, and Carbohydrates. Wiley-IEEE, Ch. A2.3 Measurement of Water Activity Using Isopiestic Method, p. 51.

Yu, L. X., Carlin, A. S., Amidon, G. L., Hussain, A. S., 2004. Feasibility studies of utilizing disk intrinsic dissolution rate to classify drugs. Int. J. Pharm. 270, 221–227.

Zakeri-Milani, P., Barzegar-Jalali, M., Azimi, M., Valizadeha, H., 2009. Biopharmaceutical classification of drugs using intrinsic dissolution rate (IDR) and rat intestinal permeability. Eur. J. Pharm. Biopharm. 73, 102–106.

Zeitler, J. A., Taday, P. F., Gordon, K. C., Pepper, M., Rades, T., 2007. Solid-state transition mechanism in carbamazepine polymorphs by time-resolved terahertz spectroscopy. ChemPhysChem 8, 1924–1927.

3.2 Effect of Crospovidone and Hydroxypropyl Cellulose on Carbamazepine in High-Dose Tablet Formulation

Abstract

The aim of this study was to develop a high-dose tablet formulation of the poorly soluble carbamazepine (CBZ) with sufficient tablet hardness and immediate drug release. A further aim was to investigate the influence of various commercial CBZ raw materials on the optimized tablet formulation. *Materials and Methods:* Hydroxypropyl cellulose (HPC-SL) was selected as a dry binder and crospovidone (CrosPVP) as a superdisintegrant. A direct compacted tablet formulation of 70% CBZ was optimized by a 3^2 full factorial design with two input variables, HPC (0–10%) and CrosPVP (0–5%). Response variables included disintegration time, amount of drug released at 15 and 60 min, and tablet hardness, all analyzed according to USP 31. *Results and Discussion:* Increasing HPC-SL together with CrosPVP not only increased tablet hardness but also reduced disintegration time. Optimal condition was achieved in the range of 5–9% HPC and 3–5% CrosPVP, where tablet properties were at least 70 N tablet hardness, less than 1 min disintegration, and within the USP requirements for drug release. Testing the optimized formulation with four different commercial CBZ samples, its variability was still observed. Nonetheless, all formulations conformed to the USP specifications. *Conclusion:* With the excipients CrosPVP and HPC-SL an immediate release tablet formulation was successfully formulated for high-dose CBZ of various commercial sources.

Keywords: Immediate Release, Direct Compaction, Superdisintegrant, CrosPVP, Dry Binder, HPC-SL, Experimental Design.

Introduction

Carbamazepine (CBZ) is a well established drug against epilepsy and trigeminal neuralgia. It is poorly soluble with dissolution controlled absorption, and hence a class II drug according to the Biopharmaceutical Classification System (Lindenberg et al., 2004). CBZ shows a history of erratic dissolution behavior with high variability between and within different marketed CBZ tablets world-wide and clinical failures are a longstanding problem (Wang et al., 1993; Meyer et al., 1992, 1998; Davidson, 1995; Jung et al., 1997; Lake et al., 1999; Mittapalli et al., 2008).

CBZ is a polymorphic drug with at least four polymorphic forms. Form III is the thermodynamically most stable form at ambient condition and it is also the commercially available form. There are several CBZ solvates and cocrystals, e.g., CBZ monoacetonate and CBZ:saccharin cocrystal (Fleischman et al., 2003; Hickey et al., 2007). At high relative humidity and in water CBZ transforms to its slower dissolving CBZ dihydrate form. This transformation critically influences dissolution and bioavailability of CBZ formulations and it has been in the focus of many investigations over the last 20 years. Laine et al. (1984) described the transformation as solution-mediated process and the formation of dihydrate crystals by whisker growth. A decrease in dissolution rate is expected as soon as the less soluble CBZ dihydrate is formed. This phenomenon is detected as an inflection point in the disc intrinsic dissolution rate (DIDR) profile of CBZ samples. Kobayashi et al. (2000) found that the transformation depends on the polymorphic form. Šehić et al. (2010), however, reported different transformation points within commercial CBZ samples, though same polymorphic form was specified. This difference also shows in tablet formulation using Ludipress® as tablet filler. Building on the method of Šehić et al. (2010) a unidirectional dissolution method was developed to characterize DIDR of CBZ samples over a longer time range describing the transformation as a process with onset (first inflection point) and equilibrium stage of transformation (Flicker et al., 2011). The unidirectional dissolution method was further applied to study the initial drug release of CBZ of binary mixtures with commonly used tablet fillers. The water-soluble mannitol increases initial drug release of CBZ samples up to 10-fold in mixtures of 30% drug load and the presence of MCC results in reduced variability in drug release. DIDR method is presented as a straightforward monitoring tool to characterize variability of CBZ raw materials and effect of tablet fillers.

There have been several attempts to control transformation of CBZ anhydrous to dihydrate. Dihydrate formation is strongly inhibited in solutions (1–4% w/v) of hydroxypropyl methylcellulose (HPMC), hydroxypropyl cellulose (HPC), or polyvinyl pyrrolidone (PVP), whereas solutions of polyvinyl alcohol (PVA), polyethylene glycol (PEG), and sodium carboxymethyl cellulose (CMC) have only weak inhibitory activity (Gift et al., 2008; Tian et al., 2006; Otsuka et al., 2000). In contrast, surfactants, such as sodium laurylsulphate and sodium taurocholate promote the dihydrate formation during the dissolution test (Rodrìguez-Hornedo and Murphy, 2004). Few authors studied the effect on transformation by including the excipient directly into the tablet formulation. Katzhendler et al. (1998, 2000) analyzed the gel layer of matrix tablets containing HPMC and later also egg albumin during the dissolution process. HPMC was found to inhibit the transformation to the CBZ dihydrate form, however, HPMC also participates in its crystallization process and induces amorphous CBZ. Egg albumin was also found to inhibit the transformation, though with a dose dependence. Salameh and Taylor (2006) studied the transformation exposing physical mixtures of anhydrous CBZ or theophylline with mannitol, microcrystalline cellulose (MCC), PVP K12, or PVP K90 to various relative humidity. MCC has only minimal effect on both drugs, mannitol enhances dehydration, whereas PVP K12 and PVP K90 show contradictory results for CBZ and theophylline. The authors concluded, that the effect of excipients on hydrate forming drugs depends on a multitude of factors and knowledge is still limited. Recently, Schulz et al. (2010) successfully adsorbed CBZ to crospovidone to prevent recrystallization to CBZ dihydrate during the solvent disposition. However, transformation is prevented only in small drug loads ($\leq 9.1\%$).

Information about excipients used in marketed CBZ tablet formulations is sparse. Lake et al. (1999) published in their studies the excipients of four marketed tablet formulations. Not surprisingly, all formulations contained excipients (starch or cellulose derivates) enhancing the disintegration properties.

The excipients crospovidone (CrosPVP) and HPC are both promising to control the dissolution behavior to be more homogeneous as they are both reported to inhibit nucleation of the CBZ dihydrate (Otsuka et al., 2000; Tian et al., 2006; Schulz et al., 2010). CrosPVP is a water-insoluble and non-gelling polymer used as tablet disintegrant and solubility enhancer (2–5%). It is free-flowing and suitable for direct compaction (Rowe et al., 2006). Mixing CBZ with CrosPVP in a drug:polymer weight ratio of 1:2 results in drastically increased dissolution rate of the drug (Machiste et al., 1995). The excipient HPC is a water-soluble polymer widely used in tablet formulation. Depending on the substitution level and on the particle size, HPC is effective as tablet binder or as tablet disintegrant. At low substitution and as fine powders, HPC shows low viscosity and sustained release and is suitable for direct compression (Rowe et al., 2006; Kawashima et al., 1993). So far, the combination of the excipients CrosPVP and HPC has not been studied.

In this study the effect of the two excipients CrosPVP and HPC on CBZ high-dose tablet formulation were investigated by experimental design. In a second step, CBZ raw materials of different commercial sources were tested with the optimized tablet formulation for immediate release.

Materials and Methods

Materials

Carbamazepine (CBZ) was obtained from four different commercial suppliers; in this study they were designated as CBZ A, B, C, and D. *Crospovidone* (CrosPVP) a cross-linked polyvinylpyrrolidone (Polyplastone® XL-10, ISP Corp., USA) was used as the superdisintegrant. *Hydroxypropylcellulose* (HPC) of special low viscosity and fine powder type (HPC-SL, Nisso, Japan) was used as the dry binder. *Microcrystalline cellulose* MCC SANAQ® 102L was donated by Pharmatrans SANAQ AG, Switzerland. All samples were stored at room temperature (20–25°C) and controlled relative humidity (43% RH). All other chemicals and reagents were of analytical grade.

Experimental Design

The effect of CrosPVP and HPC on CBZ tablet formulation was studied by a 3^2 full factorial design. Input variables were HPC (0–10%) and CrosPVP (0–5%) at the levels minimal (−), mid-point (0), and maximal (+) concentration according to the design. Response variables included disintegration time, amount of drug released at 15 and 60 min, and tablet hardness. The software STAVEX® 5.0 (Aicos, Switzerland) was used for generating the experimental design, modeling the response surface plots, and calculating the statistical evaluation.

Preparation of Carbamazepine Tablet Formulation

Powder Blends

All powder blends were of 70% CBZ and MCC was selected as the tablet filler, as suggested in previous studies (Flicker et al., 2011). CrosPVP and HPC were added according to the experimental design. Powder blends were mixed for ten minutes with a Turbula® type T2A mixer (Willy A. Bachofen, Switzerland) and for further two minutes after adding 1% magnesium stearate (Sandoz, Switzerland). For the design of experiments and the validation of the experimental design, only CBZ of supplier C was used.

Tablet Formulation

A direct compaction method was selected to prepare tablet formulations. Tablets of 400 mg were prepared by a compaction simulator (Presster®, Metropolitan Computing Co., USA) using a 10-mm flat-faced punch. The rotary press Korsch 336 with 36 stations was simulated at a speed of 0.5 m/s. Compaction force was between 6–10 kN for all tablets, with the exception of tablets containing CBZ A (14–17 kN).

Porosity was kept at $11.9 \pm 0.3\%$. Porosity (ϵ) was calculated by the Equation (3.3) based on tablet volume,

$$\epsilon = (1 - \frac{m}{V_t \rho_t})100 \qquad (3.3)$$

where m is the tablet mass and V_t the tablet volume, and ρ_t the true density of the powder blends. Density of the powder blends was calculated as weighted mean from the true density of each component. True density was assessed by a gas displacement pycnometer (AccuPyc 1330, Micromeritics, USA).

Disintegration Test

Disintegration time was measured with a disintegration apparatus, without disk, (Sotax*DT2*, SOTAX AG, Switzerland). Disintegration medium was distilled water at $37°C \pm 2°C$.

Dissolution Test

Dissolution test was performed by a USP Apparatus II (Sotax*AT7smart*, SOTAX AG, Switzerland). The conditions were according to the USP monograph for CBZ immediate release tablets. Paddle speed was 75 rpm and media was 900 ml water containing 1% sodium laurylsulfate (w/V) at 37°C. Samples were analyzed by UV-VIS Spectrophotometer (Lambda 25, PerkinElmer) at 287 nm. Dissolution curves of CBZ tablets were evaluated according to the criteria for immediate release CBZ tablets described in the USP monograph. CBZ tablets are within specification, if 45–75% CBZ is released after 15 min (t_{15}) and not less than 75% after 60 min dissolution (t_{60}).

Tablet Hardness

Breaking strength was measured by a tablet hardness tester (Dr. Schleuniger® model 8M, Pharmatron, USA). Tensile strength (σ) was calculated by the Equation (4.1),

$$\sigma = \frac{2F}{\pi Dt} \tag{3.4}$$

where F is the force needed to fracture the tablet by the hardness tester, D the diameter, and t the thickness of the cylindrical tablet.

Interaction Studies of CBZ Samples in Powder Blends

All powder blends and excipients were analyzed by X-ray powder diffractometry (XRPD) using a diffractometer (D5000, Siemens, Germany). The powder was filled into special holders and the surface was pressed flat. Operating conditions were Ni filtered Cu-Kα radiation (λ =1.5406), 40 kV, 30 mA, 0.02° 2θ, 1.0 s, 1° 2θ/min, and angular range was 5°–40° 2θ.

Thermal behavior of all powder blends and excipients were analyzed by differential scanning calorimetry (DSC), using heat flux DSC (4000, PerkinElmer, USA). DSC was calibrated with indium prior to the measurement. A sample of 4–6 mg was accurately weighed into an aluminum pan with holes and scanned between 40°C and 220°C at 10°C/min under dry nitrogen gas purge (20 ml/min).

To ensure a stable condition for the anhydrous CBZ, not to transform to its dihydrate form, water activity of all powder blends was monitored with the digital water activity analyzer (Hygropalm, Rotronic AG, Switzerland) at 23.5 ± 1.5°C.

Friability Test

Friability of CBZ tablets with different CBZ raw material was measured by a USP conform friability tester (Erweka® type TAD-UZ, Apparatebau GmbH, Germany). Each 17 tablets, weighing close to 6.5 g were tested. The drum was rotated 100 times at 25 rpm and the weight loss (%) was recorded.

Statistical Analysis

Mean dissolution curves of the tablets with different CBZ raw materials were compared by difference factor (f_1) and similarity factor (f_2), as of Equations (3.5) and (3.6),

$$f_1 = \left\{ \frac{\sum_{t=1}^{n} |R_t - T_t|}{\sum_{t=1}^{n} R_t} \right\} \times 100 \qquad (3.5)$$

$$f_2 = 50 \times \log \left\{ \left(1 + \frac{1}{n} \sum_{t=1}^{n} (R_t - T_t)^2 \right)^{-0.5} \times 100 \right) \qquad (3.6)$$

where n is the number of dissolution samplings, and R_t, T_t are the percent dissolved of the reference and test dissolution profile at each time point t.

Mean dissolution curves were further compared by Turkey test. At time points t_{15} and t_{60} individual amount of drug released was compared using one-way ANOVA.

Results and Discussion

Experimental Design

The results of response variable to the 3^2 full factorial design are displayed in Table 3.5. The responses disintegration time and amount of drug dissolved varied strongly. Disintegration time varied from 0.18 min (11 sec) to 240 min. The amount of drug dissolved was 2-75% and 7-92% after 15 and 60 min, respectively (Figure 3.10). This is expected as the factors covered a wide concentration range. The amount of drug dissolved was 2–75% and 7–92% after 15 and 60 min, respectively (). The tablet hardness was 63–80 N in breaking strength corresponding to 0.970–1.193 tensile strength.

The response variables could be described by the following equations, where D and B are the factors disintegrant CrosPVP and binder HPC. Equation (3.7) describes the disintegration time by a logarithmic function [log(DT)], Equations (3.8) and (3.9) describe the amount of drug dissolved at 15 and 60 min dissolution time (dd_{15}, dd_{60}), and Equation (3.10) the breaking strength (BS). Table 3.6 shows the goodness of fit (R^2) and adjusted goodness of fit (Rc^2) for the equations. All equations showed good correlation for the experimental data ($Rc^2 > 0.840$).

$$log(DT) = 3.90 - 2.78*D + 0.028*B + 0.32*D^2 + 0.01*B^2 - 0.01*D*B \quad (3.7)$$

$$dd15 = 14.24 + 28.52*D + 0.35*B - 3.18*D^2 - 0.23*B^2 - 0.30*D*B \quad (3.8)$$

$$dd60 = 24.74 + 36.83*D - 0.17*B - 4.69*D^2 - 0.19*B^2 + 0.09*D*B \quad (3.9)$$

$$BS = 68.89 + 1.87*D - 1.90*B - 0.05*D^2 + 0.25*B^2 - 0.08*D*B \quad (3.10)$$

Table 3.5: The 3^2 full factorial design for the factors CrosPVP (cPVP) and HPC at low (−), mid-point (0), and high (+) level concentration and its responses disintegration time (DT), drug dissolved after 15 and 60 min (dd_{15}, dd_{60}), and breaking strength (BS).

	levels		% (w/w)		response variables			
Nr.	cPVP	HPC	cPVP	HPC	DT [min]	dd_{15}[%]	dd_{60}[%]	BS [N]
1	−	−	0	0	35.00	12.4 ± 1.4	25.7 ± 1.6	69 ± 1
2	0	−	2.5	0	0.83	71.0 ± 3.5	87.8 ± 4.8	74 ± 2
3	+	−	5	0	0.18	73.9 ± 1.3	90.3 ± 1.9	76 ± 3
4	−	0	0	5	105.00	4.6 ± 0.2	14.5 ± 0.6	65 ± 2
5	0	0	2.5	5	0.43	66.1 ± 7.6	88.1 ± 5.6	68 ± 1
6	+	0	5	5	0.33	63.4 ± 4.0	88.1 ± 2.3	73 ± 1
7	−	+	0	10	240.00	2.5 ± 0.3	8.1 ± 0.8	75 ± 2
8	0	+	2.5	10	1.40	25.5 ± 6.6	64.1 ± 5.1	77 ± 2
9	+	+	5	10	0.82	49.1 ± 7.7	77.3 ± 5.5	78 ± 2

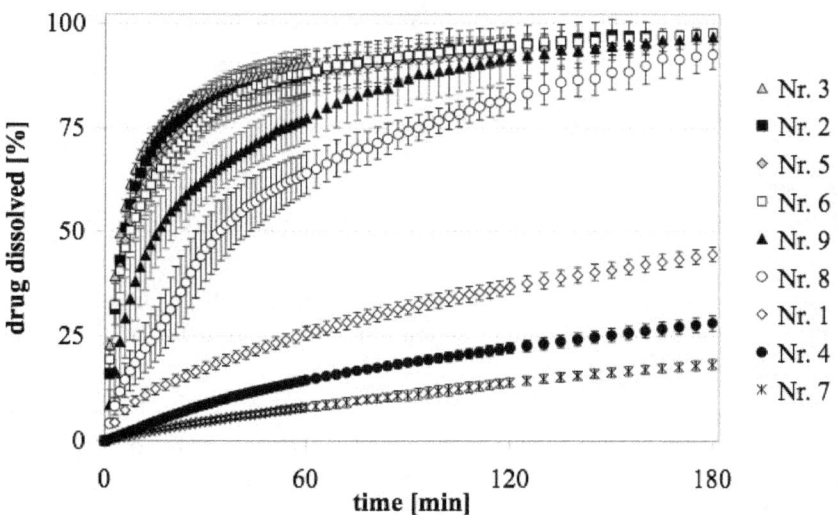

Figure 3.10: Dissolution curves of CBZ formulations Stavex Nr. 1–9

Table 3.6: Goodness of fit for the response equations of disintegration time (DT), drug dissolved after 15 and 60 min (dd_{15}, dd_{60}), and breaking strength (BS).

responses	Equation	Goodness of fit R^2	Adj. goodness of fit Rc^2
log(DT)	(3.7)	0.9845	0.9595
dd15	(3.8)	0.9401	0.8403
dd60	(3.9)	0.9901	0.9736
BS	(3.10)	0.9680	0.9146

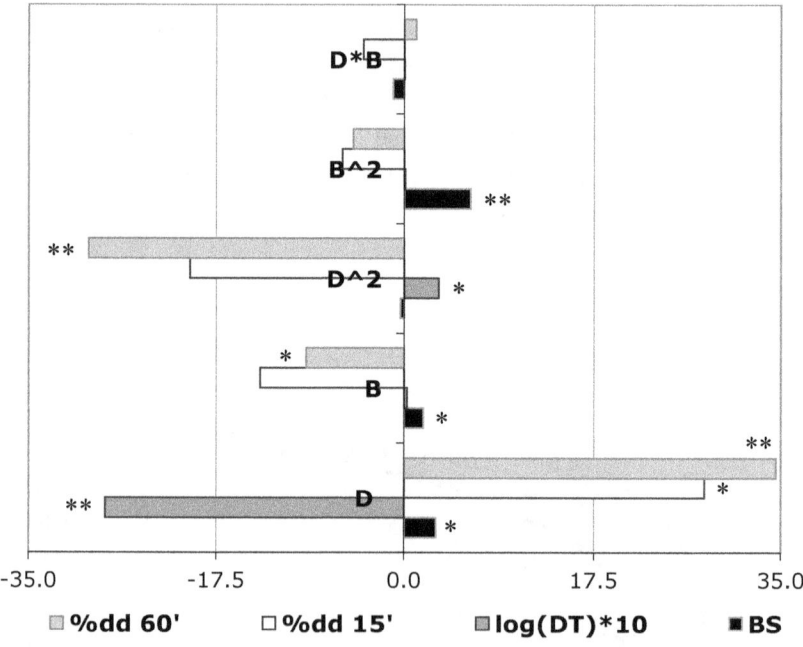

Figure 3.11: Coefficient values (coded parameters) to each factor showing the positive and negative impact on the response variables: ($*$) $p < 0.05$, ($**$) $p < 0.01$.

The impact of each factor on the response variables is shown in the pareto chart in Figure 3.11. The coefficient values for each factor by itself (D, B), concentration dependent (D^2, B^2), and in interaction (D∗B) are displayed as coded units and labeled with the significance of the factor.

Effect of CrosPVP and HPC

The effect of each factor on the tablet formulation was studied. Effect of increasing amount of CrosPVP was a shorter disintegration and faster drug release. From the model of experimental design, the minimal disintegration time was for tablet formulation of 3.8% CrosPVP, the maximal drug release at 15 and 60 minutes were found for tablets with 4.6% and 3.9% CrosPVP, and maximal tablet hardness was achieved with 5% CrosPVP.

Presence of HPC resulted in increased disintegration time and slower dissolution, and mostly in stronger tablets. Adding HPC alone improved compactability, however, drug release was prolonged. Tablet hardness is critical in CBZ formulation of high drug load. CBZ is poorly compactable and is prone to capping and lamination (Lefebvre et al., 1986; Nokhodchi et al., 2007).

Combined Effect of CrosPVP and HPC

The combined effect of CrosPVP and HPC is shown in Figure 3.12. The shortest disintegration time was achieved with medium to high amount of CrosPVP combined with same or up to the double amount of HPC. However, with this optimal amount of CrosPVP, the dissolution was faster with low to medium amounts of HPC and tablet hardness was best with either low or high amount of HPC.

Design Space for Optimal Formulation

Figure 3.13 shows the contour plots of all responses overlapped to one figure. It presents a rough visualization of the design space. The light grey window depicts the region of the optimal formulation with disintegration time of less than 1 min, drug dissolved of 45–75% at 15 min and more than 75% at 60 min dissolution time, and of sufficient breaking strength (more than 70 N). The window is confined by insufficient breaking strength and prolonged disintegration time from the left, excess in drug released after 15 min from the bottom-right, and by insufficient drug released after 60 min from bottom-left and top-right. The optimal conditions were achieved with approximately 3–5% CrosPVP and 5–9% HPC. The amount of CrosPVP is within the concentration range suggested in literature (2–5%). However, higher amount of HPC was necessary to achieve optimal formulation than the suggested 2–6% HPC as a dry binder. Nonetheless, the optimal amount of HPC is still lower than the suggested amount of 15–35% HPC for extended release formulations (Rowe et al., 2006).

Model Validation

The result of the validation is displayed in Table 3.7. The responses of three additional formulations within the design space for optimal formulation were tested

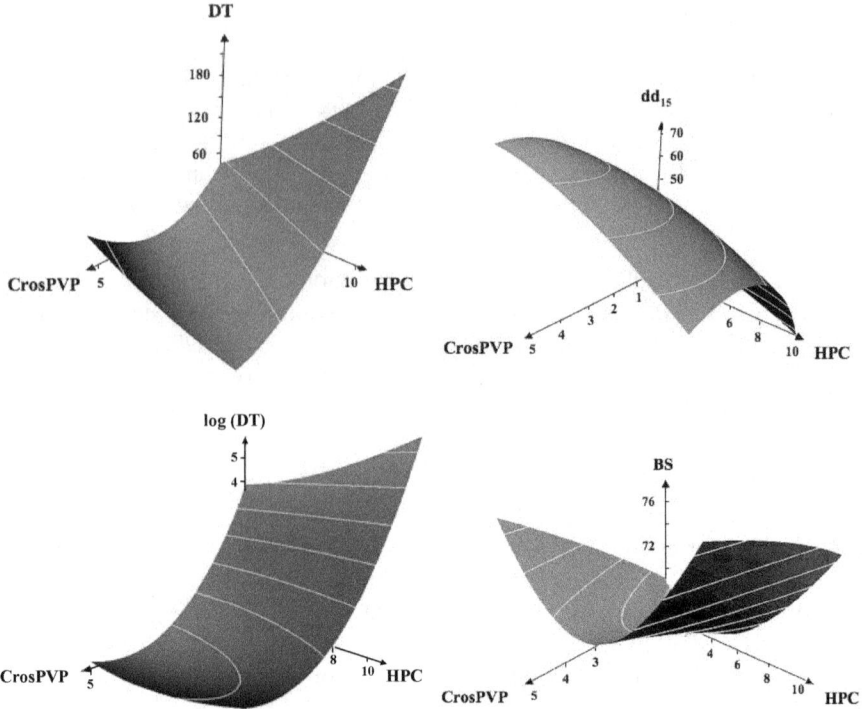

Figure 3.12: Effect of interaction in the responses disintegration time (DT), drug dissolved after 15 and 60 min (dd_{15}, dd_{60}), and breaking strength (BS)

and the deviation from the predicted responses was calculated. Formulation x (4.5% CrosPVP and 7.0% HPC) and z (5% CrosPVP and 6% HPC) were selected to test the upper region of the optimal window and formulation y (3% CrosPVP and 1% HPC) to test the lower region. The response variables were predicted with deviations of 14% and less. Only the disintegration time could not be predicted for the short disintegrations of less than 1 min (deviation > 65%).

The dissolution curves of test formulations x, y, and z are shown in Figure 3.14. All test formulations were within the USP specifications for CBZ immediate release tablets. Formulation z was selected as the optimal formulation because its dissolution curve was with 53.6% drug dissolved at t_{15} in the middle between the range required by USP (dd_{15} 45–75%) and it showed the smallest relative standard deviation within the first 60 min of dissolution (RSD_{60} 3.8%). Furthermore, tablets were with 79.9 N of sufficient hardness.

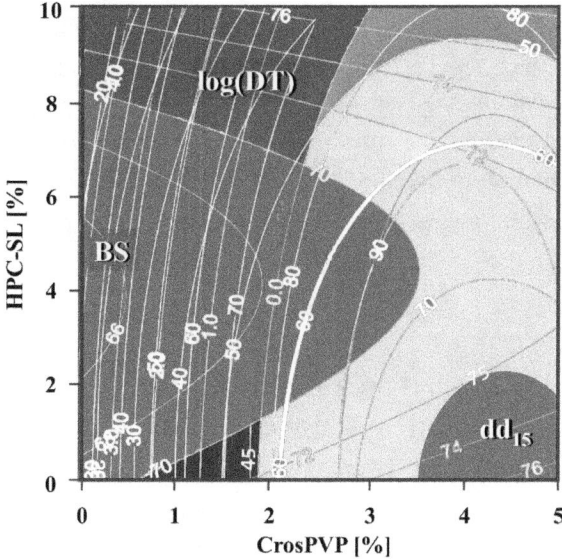

Figure 3.13: Contour plots of response variables log(DT), dd_{15}, dd_{60}, and BS, overlapped to one figure, roughly visualizing the region for the design space for the optimal formulation (light grey window).

Observations on Disintegration Behavior and Dissolution Variability

Tablet formulations showed a remarkable wide range in RSD_{60} that could be attributed to the effect of CrosPVP and HPC-SL. Dissolution curves of tablets from experiment number 3 showed a RSD_{60} of less than 2%, dissolution curves of tablets from experiment numbers 7–9 showed a RSD_{60} of more than 12%. Analyzing RSD_{60} as an additional response variable to the experimental design, a good correlation with the input variables was found (Rc^2 0.9087). CrosPVP showed a positive effect on the RSD_{60}, where increasing amount of CrosPVP reduced the RSD_{60} ($p < 0.05$). HPC-SL had a negative effect on RSD_{60} with a maximal RSD_{60} at the highest amount. The reason for the negative effect of HPC-SL may be due to its function as binder and its morphological structure. Tablet formulations containing high amounts of HPC-SL formed sticky, irregular, and flake-like agglomerates, whereas tablets containing high amounts of CrosPVP presented fine particles. These effects were best visible during the disintegration and dissolution testing. The compact presenting the irregular and flake-like agglomerates also showed the highest RSD_{60}.

Table 3.7: Model validation for the response variables disintegration time (DT), drug dissolved after 15 and 60 min (dd_{15} and dd_{60}), and breaking strengthe (BS), for the test formulations x, y, and z.

response	predicted	observed	% deviation
formulation x: 4.5% CrosPVP, 7% HPC			
log(DT) [min]	0.7[1]	0.5	65.4
dd_{15} [%]	60.1	58.4	2.9
dd_{60} [%]	88.0	87.0	1.2
BS [N]	76.1[1]	84.6	11.2
formulation y: 3% CrosPVP, 1% HPC			
log(DT) [min]	-0.6[1]	0.32	68.1
dd_{15} [%]	71.5	73.0	2.1
dd_{60} [%]	92.9	88.1	-5.2
BS [N]	72.8[1]	76.0	4.4
formulation z: 5% CrosPVP, 6% HPC			
log(DT) [min]	0.7[1]	0.4	66.1
dd_{15} [%]	62.3	53.6	14.0
dd_{60} [%]	86.6	84.0	3.0
BS [N]	75.7[1]	79.7	5.4

[1] predicted by partial equation, $p < 0.1$

Optimized Tablet Formulation of CBZ of Different Suppliers

CBZ samples of four different suppliers were tested with the formulation z. As the effect of CrosPVP and HPC-SL on drug release variability was observed in the dissolution test, the influence of total amount of these excipients was tested for its ability to reduce the variability found for the different CBZ samples. A formulation of high level CrosPVP and HPC-SL was selected (formulations CBZ_h: 7.5% CrosPVP and 9% HPC-SL). The ratio of CrosPVP to HPC-SL was the same as in the optimal formulation z, but out of the design space. As a counter point a formulation of low level CrosPVP and HPC-SL was selected within the design space (formulations CBZ_l: 2% CrosPVP and 1% HPC-SL). The dissolution curves of the formulations CBZ_l and CBZ_h are shown in Figure 3.15 together with formulation z, each with CBZ A, B, C, or D.

The dissolution curves were compared by the similarity and difference factor (f_1, f_2). The results are shown in Table 3.8. Dissolution curves of tablets with CBZ C were used as the reference. According to the FDA guidance for industry (FDA, 1997) sameness or equivalence of two curves is assumed if f_1 values are 0–15 and f_2

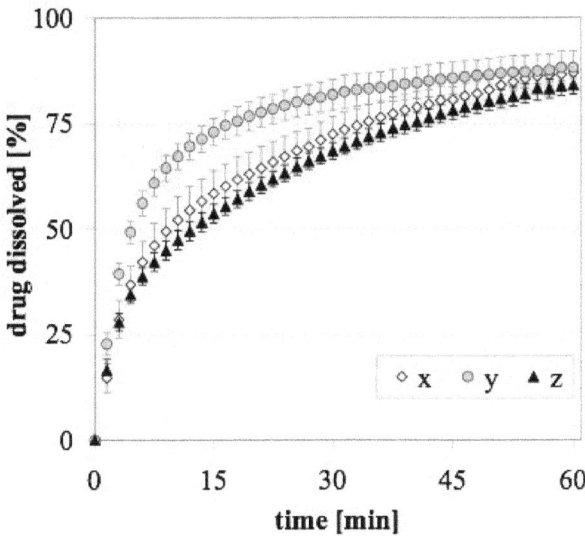

Figure 3.14: Dissolution curves of test formulations x, y, and z.

values 50–100. Applying this rule to the results, dissolution curves were the same among tablets of CBZ A, B, C, and D at all three concentration levels, with the exception of the dissimilar dissolution curves of tablets containing CBZ B and C at high total amount of excipients. Dissolution curves of tablets containing CBZ A and C were with f_1 of 15 and f_2 of 50 at the upper limit to still conclude sameness.

In contrast, pairwise comparison by Turkey Test (95% CI) showed significant difference among each sample at all concentration levels ($p < 0.01$), except between low level tablets of CBZ A and B, and between low and high level tablets of CBZ C and D. Comparing amount of drug dissolved at one time point (15 and 60 min), the difference among tablets with CBZ A, B, C, and D was even more pronounced (ANOVA $p < 0.005$). Also Yuksel et al. (2000) observed the ANOVA-based methods to be more discriminative and more informative on differences in dissolution curves than the f-factors.

It has to be noted, that the dissolution curves of the formulations with high excipient levels showed a RSD_{60} of more than 10%. At low excipient level dissolution curves showed also elevated RSD_{60}, but less than 8%. The high RSD_{60} obscured the different dissolution behavior of each CBZ raw material. Increasing or decreasing the total amount of CrosPVP and HPC-SL could not reduce the influence of raw drug variability on the dissolution behavior, proving the optimal concentration of 5% CrosPVP and 6% HPC-SL in formulation z.

As to the fast disintegration of less than 1 min, the inhibitory effect of HPC and CrosPVP on the CBZ transformation could not be detected during the dissolution. The CBZ samples used in this study were reported to show individual transforma-

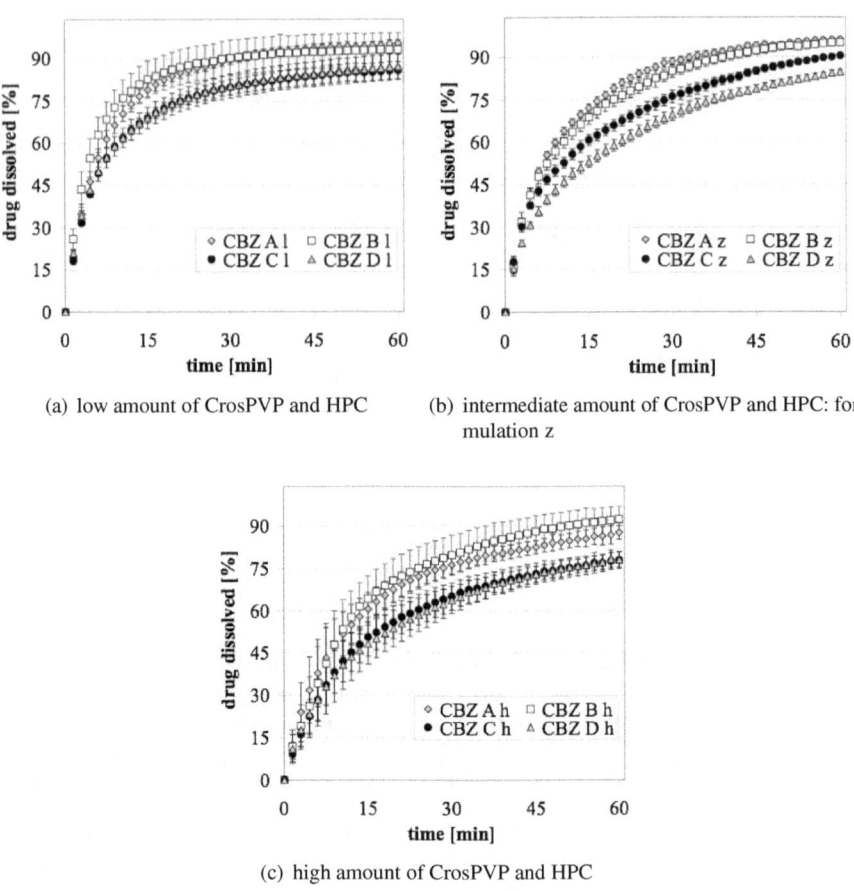

Figure 3.15: Dissolution curves with CBZ A, B, C, and D of low (l), intermediate (z), and high (h) concentration of CrosPVP and HPC.

tion behavior in previous work and MCC is suggested as a tablet filler to reduce variability in drug release (Flicker et al., 2011). Furthermore, CrosPVP and HPC were also reported to reduce variability in drug release, because of their ability to inhibit transformation. However, the results of the current study still show variability in CBZ tablet formulation according to the CBZ source used.

To verify the reason for this variability, the samples could be analyzed by in-situ Raman spectroscopy during the dissolution process (Tian et al., 2007; Savolainen et al., 2009). Besides an individual transformation behavior, morphological differences can not be excluded as a reason to the differences among tablets of different CBZ source.

Table 3.8: Similarity and difference factor (f_1, f_2) for dissolution curves of tablets with CBZ A, B, C, and D at low (CBZ_l), intermediate (CBZ_z), and high (CBZ_h) total concentration of CrosPVP and HPC.

		f_1; f_2	f_1; f_2	f_1; f_2
		A	B	D
CBZ_l	C	11 ; 52	12 ; 50	1 ; 94
CBZ_z	C	11 ; 52	9 ; 58	8 ; 61
CBZ_h	C	15 ; 50	20 ; 43	1 ; 91

Tablet Hardness and Friability

Tablets of the optimized formulation z showed different hardness depending on the CBZ sample. Breaking strength was 150 ± 1 N, 79 ± 2 N, 80 ± 2 N, and 81 ± 2 N for tablets with CBZ A, B, C, and D, respectively.

Friability was 0.5%, 1.2%, 1.3%, and 0.9% for the optimized formulation z with CBZ A, B, C, and D, respectively. Only the formulation with CBZ A and D showed enough robustness for possible film coating. The variability of tablet hardness and friability regarding CBZ raw materials emphasizes a concern of appropriate selection of raw material supply.

Interaction Studies in Powder Blends

The results for XRPD and DSC analysis of the powder blends are shown in Figure 3.16. No interaction or change in polymorphic form was detected.

For all powder blends water activity was between 0.435 and 0.528. Phase conversion of CBZ p-monoclinic form to CBZ dihydrate depends on the water activity of the system (Qu et al., 2006) as both forms were found to coexist in equilibrium at the water activity range 0.601 to 0.641 and 20°C (Li et al., 2008). Our samples were therefore not in the critical range.

Conclusions

High-dose CBZ tablets were successfully prepared with the aid of CrosPVP and HPC-SL. Effect of CrosPVP and HPC on a high-dose CBZ tablet formulation were clearly visible. An optimal formulation was found containing 5% CrosPVP and 6% HPC, where tablet formulation of different CBZ raw materials conformed to the USP requirements for CBZ immediate release tablets. Nonetheless, variability in commercial CBZ samples effected the tablet properties including the dissolution

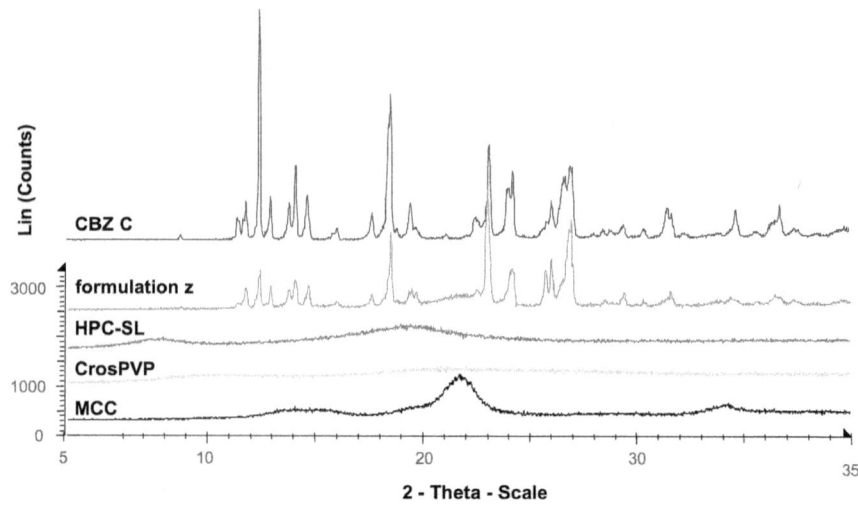

(a) XRPD: 5°–40° 2θ at 0.02° 2θ and 1° 2θ/min.

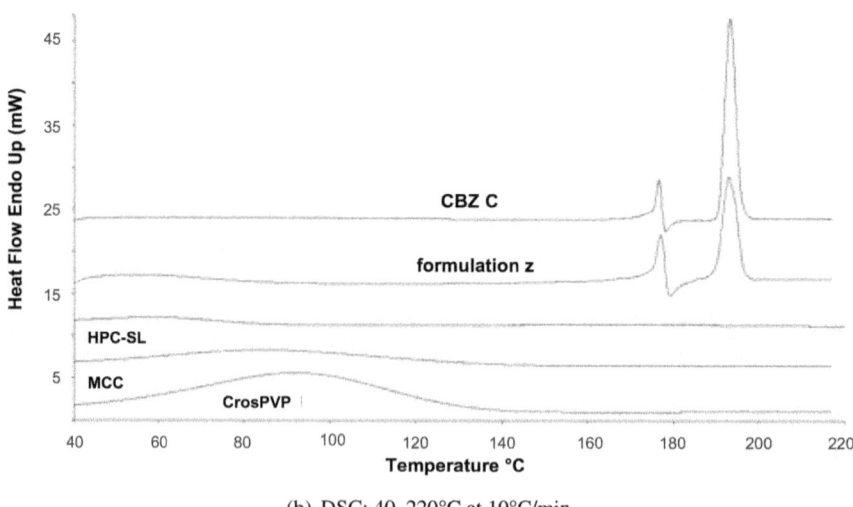

(b) DSC: 40–220°C at 10°C/min

Figure 3.16: X-ray powder diffraction (top) and DSC (bottom) of CBZ N, powder blend of formulation z, and components HPC-SL, CrosPVP, and MCC.

behavior and its influence could not be reduced. We therefore suggest selecting the CBZ raw material carefully to obtain a robust high-dose formulation.

Acknowledgments

The financial support by the Senglet Stiftung Basel, Switzerland, in form of a Ph.D. scholarship for Felicia Flicker is kindly appreciated.

Declaration of Interest

The authors report no conflicts of interest. The authors alone are responsible for the content and writing of this paper.

Bibliography

Davidson, A., 1995. A multinational survey of the quatity of carbamazepine tablets. Drug Dev. Ind. Pharm. 21 (19), 2167–2186.

FDA, 1997. Guidance for industry: dissolution testing of immediate release solid oral dosage forms. Center for drug evaluation and research, Rockville, MD August.

Fleischman, S. G., Kuduva, S. S., McMahon, J. A., Moulton, B., Walsh, R. D. B., Rodrìguez-Hornedo, N., Zaworotko, M. J., 2003. Crystal engineering of the composition of pharmaceutical phases: multiple-component crystalline solids invloving carbamazepine. Crystal Growth and Design 3 (6), 909–919.

Flicker, F., Eberle, V. A., Betz, G., 2011. Variability in commercial carbamazepine samples – impact on drug release. Int. J. Pharm. 410, 99–106.

Gift, A. D., Luner, P. E., Luedeman, L., Taylor, L. S., 2008. Influence of polymeric excipients on crystal hydrate formation kinetics in aqueous slurries. J. Pharm. Sci. 97 (12), 5198–211.

Hickey, M. B., Peterson, M. L., Scoppettuolo, L. A., Morrisette, S. L., Vetter, A., Guzmán, H., Remenar, J. F., Zhang, Z., Tawa, M. D., Haley, S., Zaworotko, M. J., Örn Almarsson, 2007. Performance comparison of a co-crystal of carbamazepine with marketed product. Eur. J. Pharm. Biopharm. 67, 112–119.

Jung, H., Milán, R., Girard, M., León, F., Montoya, M., 1997. Bioequivalence study of carbamazepine tablets: in vitro/in vivo correlation. Int. J. Pharm. 152, 37–44.

Katzhendler, I., Azoury, R., Friedman, M., 1998. Crystalline properties of carbamazepine in sustained release hydrophilic matrix tablets based on hydroxypropyl methylcellulose. J. Control. Release 54, 69–85.

Katzhendler, I., Azoury, R., Friedman, M., 2000. The effect of egg albumin on the crystalline properties of carbamazepine in sustained release hydrophilic matrix tablets and in aqueous solutions. J. Control. Release 65, 331–343.

Kawashima, Y., Takeuchi, H., Hino, T., Niwa, T., Lin, T.-L., Sekigawa, F., Kawahara, K., 1993. Low-substituted hydroxypropylcellulose as a sustained-drug release matrix base or disintegrant depending on its particle size and loading in formulation. Pharmaceutical Research 10, 351–355.

Kobayashi, Y., Ito, S., Itai, S., Yamamoto, K., 2000. Physicochemical properties and bioavailability of carbamazepine polymorphs and dihydrate. Int. J. Pharm. 193, 137–146.

Laine, E., Tuominen, V., Ilvessalo, P., Kahela, P., 1984. Formation of dihydrate from carbamazepine anhydrate in aqueous conditions. Int. J. Pharm. 20, 307–314.

Lake, O. A., Olling, M., Barends, D. M., 1999. In vitro/in vivo correlations of dissolution data of carbamazepine immediate release tablets with pharmacokinetic data obtained in healthy volunteers. Eur. J. Pharm. Biopharm. 48 (1), 13–19.

Lefebvre, C., Guyot-Hermann, A. M., Draguet-Brughmans, M., Bouch, R., Guyot, J. C., 1986. Polymorphic transitions of carbamazepine during grinding and compression. Drug Dev. Ind. Pharm. 12, 1913–1927.

Li, Y., Chow, P. S., Tan, R. B. H., Black, S. N., 2008. Effect of water activity on the transformation between hydrate and anhydrate of carbamazepine. Org. Process Res. Dev. 12, 264–270.

Lindenberg, M., Kopp, S., Dressman, J. B., 2004. Classification of orally administered drugs on the world health organization model list of essential medicines according to the biopharmaceutics classification system. Eur. J. Pharm. Biopharm. 58, 265–278.

Machiste, E. O., Giunched, P., Setti, M., Conte, U., 1995. Characterization of carbamazepine in systems containing a dissolution rate enhancer. Int. J. Pharm. 126, 65–72.

Meyer, M. C., Staughn, A. B., Mhatre, R. M., Shah, V. P., Williams, R. L., Lesko, L. J., 1998. The relative bioavailability and in vivo - in vitro correlations for four marketed carbamazepine tablets. Pharm. Res. 15 (11), 1787–1791.

Meyer, M. C., Straughn, A. B., Jarvi, E. J., Wood, G. C., Pelsor, F. R., Shah, V. P., 1992. The bioinequivalence of carbamazepine tablets with a history of clinical failures. Pharm. Res. 9, 1612–1616.

Mittapalli, P. K., Suresh, B., Hussaini, S. S. Q., Rao, Y. M., Apte, S., 2008. Comparative in vitro study of six carbamazepine products. AAPS PharmSciTech 9 (2), 357.

Nokhodchi, A., Maghsoodi, M., Hassan-Zadehb, D., Barzegar-Jalali, M., 2007. Preparation of agglomerated crystals for improving flowability and compactibility of poorly flowable and compactible drugs and excipients. Powder Technol 175, 73–81.

Otsuka, M., Ohfusa, T., Matsuda, Y., 2000. Effect of binders on polymorphic transformation kinetics of carbamazepine in aqueous solution. Colloid Surface 17, 145–152.

Qu, H., Louhi-Kultanen, M., Kallas, J., 2006. Solubility and stability of anhydrate/hydrate in solvent mixtures. Int. J. Pharm. 321, 101–107.

Rodrìguez-Hornedo, N., Murphy, D., 2004. Surfactant-facilitated crystallization of dihydrate carbamazepine during dissolution of anhydrous polymorph. J. Pharm. Sci. 93, 449–460.

Rowe, R. C., Sheskey, P. J., , Owen, S. C. (Eds.), 2006. Handbook of Pharmaceutical Excipients, Part 3, 5th Edition. Pharmaceutical Press.

Salameh, A. K., Taylor, L. S., 2006. Physical stability of crystal hydrates and their anhydrates in the presence of excipients. J. Pharm. Sci. 95, 446–461.

Savolainen, M., Kogermann, K., Heinz, A., Aaltonen, J., Peltonen, L., Strachan, C., Yliruusi, J., 2009. Better understanding of dissolution behaviour of amorphous drugs by in situ solid-state analysis using raman spectroscopy. Eur. J. Pharm. Biopharm. 71, 71–79.

Schulz, M., Fussnegger, B., Bodmeier, R., 2010. Adsorption of carbamazepine onto crospovidone to prevent drug recrystallization. Int. J. Pharm. 391, 169–176.

Tian, F., Sandler, N., Gordon, K. C., McGoverin, C. M., Reay, A., Strachan, C. J., Saville, D. J., Rades, T., 2006. Visualizing the conversion of carbamazepine in aqueous suspension with and without the presence of excipients: A single crystal study using SEM and Raman microscopy. Eur. J. Pharm. Biopharm. 64, 326–335.

Tian, F., Zhang, F., Sandler, N., Gordon, K., McGoverin, C., Strachan, C., Saville, D., Rades, T., 2007. Influence of sample characteristics on quantification of carbamazepine hydrate formation by X-ray powder diffraction and Raman spectroscopy. Eur. J. Pharm. Biopharm. 66, 466–474.

Šehić, S., Betz, G., Hadžidedić, Š., El-Arini, S. K., Leuenberger, H., 2010. Investigation of intrinsic dissolution behavior of different carbamazepine samples. Int. J. Pharm. 386, 77–90.

Wang, J., Shiu, G., Ting, O., Viswanathan, C., Skelly, J., 1993. Effects of humidity and temperature on in-vitro dissolution of carbamazepine tablets. J. Pharm. Sci. 82, 1002–1005.

Yuksel, N., KanÄśk, A. E., Baykara, T., 2000. Comparison of in vitro dissolution profiles by ANOVA-based, model-dependent and -independent methods. Int. J. Pharm. 209, 57–67.

Chapter 4

Recrystallization Project

Abstract

Physical properties of commercial CBZ samples can significantly influence drug release and thereby jeopardized bioequivalence of the final dosage form. The aim of this project was to reduce variability in commercial CBZ samples by recrystallization. CBZ samples of four different suppliers were recrystallized in ethanol solution containing 1% PVP. The obtained crystals were analyzed by disk intrinsic dissolution rate (DIDR), differential scanning calorimetry (DSC), and X-ray powder diffraction (XRPD). Recrystallized CBZ samples showed strongly reduced variability in intrinsic dissolution rate (IDR) and first inflection point in the DIDR profiles compared to the untreated CBZ samples. Furthermore, transformation process to CBZ dihydrate was inhibited; no dihydrate crystals were visible on compact surfaces after 8 h intrinsic dissolution measurement. DSC and XRPD showed no change in polymorphic form. Analyzing recrystallized CBZ in binary mixtures with MCC drug release showed again significant variability. The low yield of the recrystallized CBZ samples (53–70%) and the necessary but critical step of grinding need further improvement. Therefore, the careful selection of the CBZ raw material is suggested at the actual stage of knowledge.

Introduction

Pharmaceutical companies are facing a fast growing market of drug suppliers, raw materials can be obtained from suppliers in India and China at lower price. Physicochemical properties of raw materials can vary among drug suppliers with consequence on the drug performance in the final drug formulation.

CBZ is one example where sample variability among different suppliers effects the dissolution behavior of the tablet formulation. Šehić et al. (2010) reported different transformation to dihydrate within commercial CBZ samples, though same polymorphic form was specified. This difference also shows in tablet formulation using Ludipress® as tablet filler. Developing a CBZ tablet formulation for immediate release, Flicker and Betz (2012) selected the two excipients crospovidone and hydroxypopyl cellulose, both reported to inhibit transformation. However, tablet hardness, friability, and drug release varied with the CBZ sample. For CBZ the variability in drug release is closely linked to clinical failures as the bioavailability of CBZ is dissolution controlled and the pharmacological action is within a narrow therapeutic range (Lindenberg et al., 2004).

Physicochemical properties of raw materials depend on the method of production. The crystallization step of a drug can be performed with different solvents and additives. They result in different crystal habits and possible solvent inclusion, both leading to different solubility and dissolution behavior (Rodrìguez-Sponga et al., 2004). This phenomenon has also been studied on CBZ. Bolourtchian et al. (2001) compared CBZ samples recrystallized in ethanol and acetone under different conditions. Crystallization from ethanol or acetone results in polyhedral and thin plate-like crystals, respectively. Using a watering-out method needle-shaped crystals are obtained. The recrystallized CBZ samples show improved dissolution rate and compactibility compared to the untreated samples. However, interpreting the XRPD results of the CBZ crystals obtained by the watering-out method, the crystals are most likely of CBZ dihydrate form. The peaks around 5, 9, 12° 2θ conform with the XRPD data reported for CBZ dihydrate (McMahon et al., 1996; Li et al., 2000). Very similar studies were performed by Mahalaxmi et al. (2009), they recrystallized CBZ samples in ethanol and acetone at different cooling conditions. The solvents also lead to different crystal habit and different polymorphic form. Fast cooling results in smaller crystals with faster dissolution. Nokhodchi et al. (2005) studied the effect of the additives PEG 4000, PVP K30, and Tween 80 (1% w/v) on the recrystallization of CBZ in ethanol solutions. Thus obtained crystals are of different crystal habit but of same polymorphic form (CBZ form III). Recrystallization in presence of PVP K30 results in more block-shaped crystals showing higher dissolution rate and improved tensile strength.

Crystallization is mostly followed by a milling or grinding step. However, these physical processes cause crystal defects, mechanical activation, and small amounts of amorphous that can further lead to faster and inhomogeneous dissolution behavior of CBZ (Tian et al., 2006b; Murphy et al., 2002; Mosharraf et al., 1999; Lefebvre et al., 1986).

In this study, CBZ samples of four different suppliers were recrystallized in ethanol containing 1% PVP. The effect of recrystallization on sample variability was investigated by DSC, XRPD, and intrinsic dissolution profiles using the unidirectional dissolution method developed earlier (Flicker et al., 2011). Recrystallized samples were also tested in binary mixtures with MCC at 70% drug load.

Materials and Methods

Carbamazepine of four different suppliers (A, B, C, and D) was recrystallized according to Nokhodchi et al. (2005). Each sample (42 g) was dissolved in 910 ml 1% polyvinylpyrrolidone (PVP K30, Kollidon®K30 BASF) ethanol solution at 65 °C. After cooling to room temperature the solutions were placed in the refrigerator (4–8 °C) for 24 h. The crystals were filtered under suction, dried at ambient condition over night, and for 72 h over pentoxide (0% RH). The crystals were slightly ground with mortar and pistil and kept at room temperature and 43% RH. Recrystallized CBZ samples were obtained at a yield of 53–70%.

Microcrystalline cellulose (MCC SANAQ®102L; Pharmatrans SANAQ AG, Switzerland) was used as a water-insoluble tablet-filler for the binary mixtures. All other chemicals and reagents purchased from commercial sources were of analytical grade.

Polymorphic Characterization

Polymorphic form of CBZ samples was characterized by X-ray powder diffractometry (XRPD) using a diffractometer (D5000, Siemens, Germany). The powder was filled into special holders and the surface was pressed flat. Operating conditions were Ni filtered Cu-Kα radiation (λ =1.5406), 40 kV, and 30 mA. Step was 0.02° 2θ, step time 1.0 s, angular scanning speed 1° 2θ/min, and angular range between 5° and 40° 2θ scale.

Thermal behavior of all samples was analyzed by differential scanning calorimetry (DSC), using heat flux DSC (4000, PerkinElmer, USA). DSC was calibrated with indium prior to the measurement. A sample of 3-6 mg was accurately weighted into an aluminum pan with holes and scanned between 40 °C and 220 °C at 10 °C/min under dry nitrogen gas purge (20 ml/min).

True density of all samples was assessed by a gas displacement pycnometer (Accu-Pyc 1330, Micromeritics, USA). Powder was purged with helium by five repetitive purging cycles and the density was reported as average value. The test was performed in triplicate.

Morphologic Characterization

For morphological characterization scanning electron microscopy (SEM) (ESEM XL 30 FEG, Philips, The Netherlands) was applied at a voltage of 10 kV and magnifications of 100–2000 times. Before analysis, powder was sprinkled on carbon adhesive and then sputtered with platinum.

Water Activity Measurement

Water activity of all samples was monitored with a digital water activity analyzer (Hygropalm, Rotronic AG, Switzerland) at 23.5 ± 1.5 °C.

Unidirectional Dissolution Method

Sample preparation. Recrystallized CBZ samples were mixed with MCC for 10 min in a mixer (Turbula® type T2C, W. Bachofen, Switzerland) and for a further 2 min after adding 1% magnesium stearate (Sandoz, Switzerland) to the mixture. Drug load was 70% and mixtures of 10 g were prepared.

Flat-faced compacts of 400 mg were produced from recrystallized CBZ and of binary mixtures using Zwick 1478 material tester (Zwick, Germany). Surface area was 0.95 cm^2 and porosity was 7% and 12% for pure drug and binary mixtures, respectively. Compaction speed was at 10 mm/min and 50 mm/min.

Dissolution. To obtain disc intrinsic dissolution rate (DIDR) profiles of the CBZ samples, unidirectional dissolution was performed by a modified USP Apparatus I. The compact was placed in a sample holder fitting to the rotating unit of the dissolution apparatus (SotaxAT7*smart*, SOTAX AG, Switzerland) and the compact was embedded in melted paraffin wax so only one surface was available to the dissolution media. Unidirectional dissolution method was performed at two conditions: one for the analysis of first inflection point referring to the start of transformation, and another for the second inflection point referring to the stabilized transformation rate. Release media was 500 ml and 1000 ml water, and run time was 2 h and 10–11 h for these two conditions, respectively. Dissolution media was at 37 ± 0.5 °C and a rotation speed was 100 rpm. Drug content of the media was measured at predetermined time intervals by UV-VIS spectrophotometer (Lambda 25, PerkinElmer, USA) at 285 nm.

Evaluation of DIDR Profiles. IDR was determined by the slope in the DIDR profile prior to the first inflection point, the slope referring to the release of anhydrous CBZ before transforming to the dihydrate. DIDR profiles were analyzed by inflection points as described in earlier work (Flicker et al., 2011). For all statistical comparison, one-way ANOVA followed by Student's t-test was applied.

Analysis of Compact Surface. Compact surfaces were analyzed by SEM (see Section on Morphological Characterization). Prior to the analysis the compacts were kept under controlled RH of 43% for a maximum of 24 h. The compacts were then fixed to the sample holder by conductive silver and sputtered with gold.

Tensile strength

Tensile strength (σ) was calculated by the Equation (4.1),

$$\sigma = \frac{2F}{\pi Dt} \tag{4.1}$$

where F is the force needed to fracture the tablet by the hardness tester, D the diameter, and t the thickness of the cylindrical tablet. Tablet dimensions were measured by a digital caliper and crushing strength by a tablet hardness tester (Dr. Schleuniger® model 8M, Pharmatron, USA).

Results and Discussion

Polymorphic characterization

Diffractograms of recrystallized CBZ samples are shown in Figure 4.1. The diffractograms were consistent with XRPD data of the untreated CBZ samples and the data reported for CBZ p-monoclinic form, where main peaks show at 2θ = 14.9, 15.2, 15.8, 27.2, 27.5, and 32.0° (Rustichelli et al., 2000). Peaks at angles smaller than 10° 2θ indicating presence of CBZ triclinic (form I) or trigonal (form II) were not detected in our samples. However, at lower angles of ° 2θ diffractograms peak intensities varied indicating difference in crystallinity, particle size, and preferred orientation.

DSC measurements of recrystallized CBZ (Figure 4.2) showed the characteristic thermal events for CBZ p-monoclinic form (Grzesiak et al., 2003; Šehić et al., 2010). Melting of form III was visible around 176 °C, followed by an exotherm around 182 °C (crystallization of form I). Around 191 °C form I melted. DSC profiles of all recrystallized CBZ samples were all alike. The deviations found in untreated CBZ samples were removed by the recrystallization.

True density of all untreated and recrystallized CBZ samples were both 1.340 ± 0.001 g/cm^3 indicating CBZ polymorphic form III (p-monoclinic) (Krahn and Mielck, 1989; Grzesiak et al., 2003).

Morphological Characterization

Figure 4.3 shows the SEM images of recrystallized CBZ samples. CBZ crystals obtained after recrystallization was very big and agglomerated at the bottom of the beaker. Prior to the analysis they were broken apart and reduced in size by mortar and pistil. The SEM images illustrate the inhomogeneous particle size and shape in the recrystallized CBZ samples.

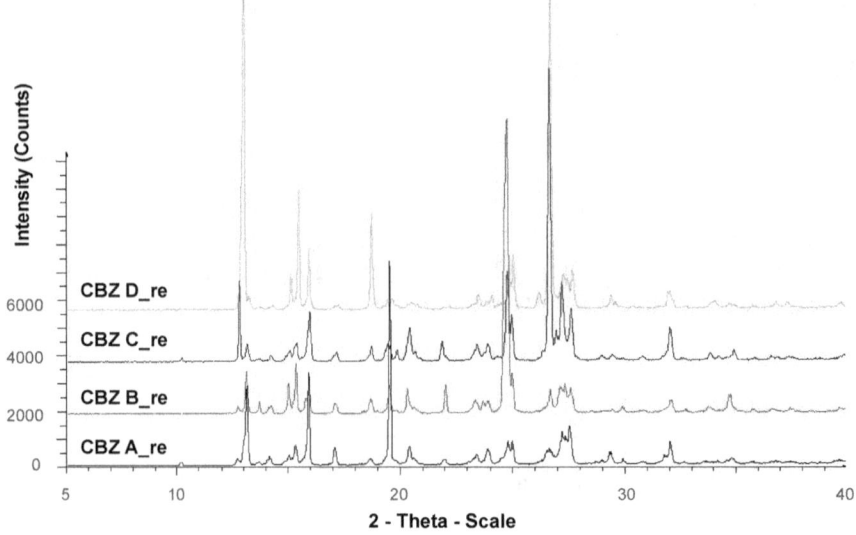

Figure 4.1: X-ray powder diffraction of recrystallized CBZ samples.

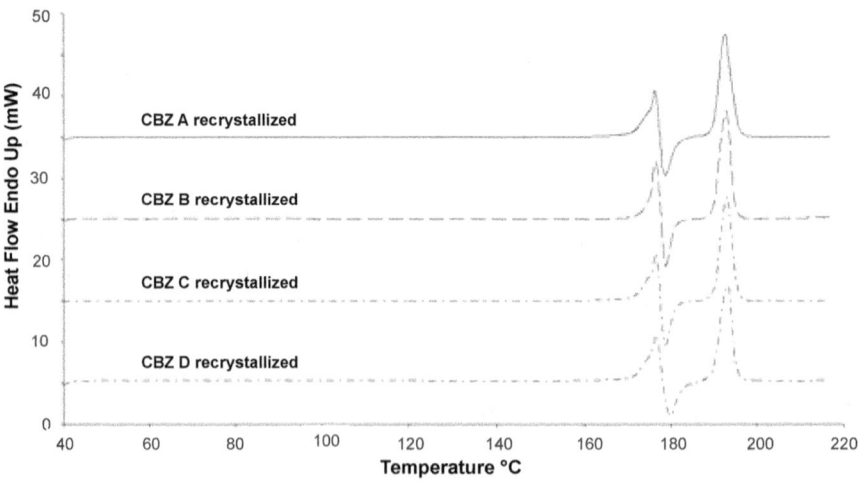

Figure 4.2: DSC profiles of recrystallized CBZ samples; 10 °C/min scanning rate.

Water activity

Water activity of recrystallized CBZ samples was 0.327–0.346 and therefore below the critical for transformation to CBZ dihydrate.

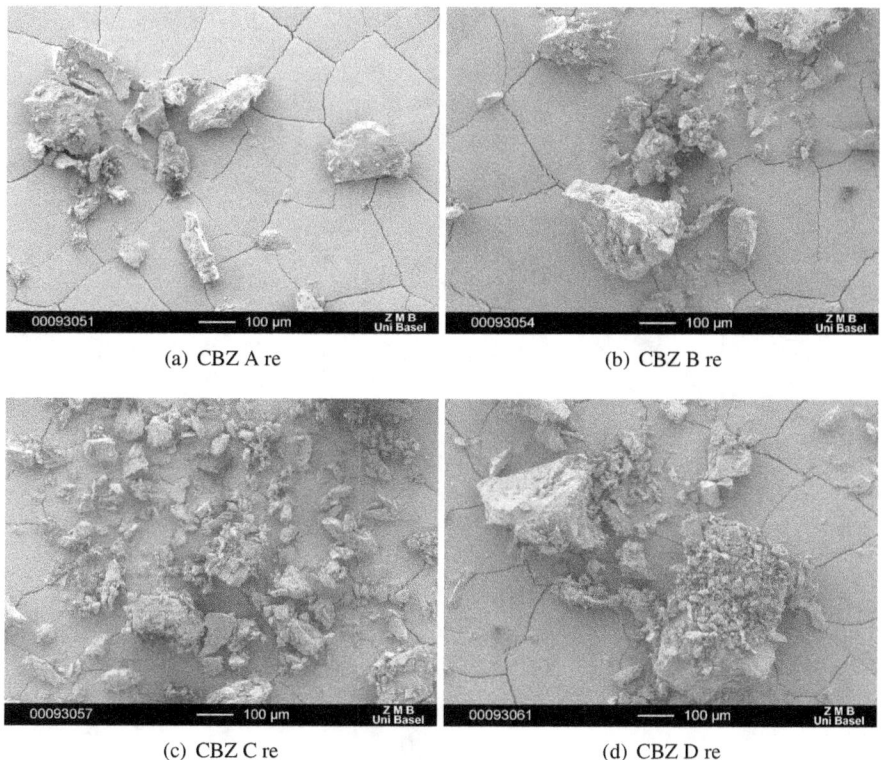

Figure 4.3: SEM pictures of recrystallized CBZ samples.

DIDR profiles of recrystallized CBZ

DIDR profiles of recrystallized CBZ A, B, C, and D are shown in Figure 4.4. Compared to the DIDR profiles of untreated CBZ samples (Section 3.1) the variability was clearly reduced. IDR values of recrystallized CBZ were with 35.3 ± 2.3 µg/cm^2/min slightly lower than the average IDR of untreated CBZ samples and with less variability. The average IDR values of untreated CBZ was 36.6 ± 8.5 µg/cm^2/min.

Inflection points in DIDR profiles of recrystallized CBZ samples are shown in Table 4.1. The values were not distinct (one-way ANOVA, p > 0.05). Furthermore, SEM images did not show any needle-like dihydrate structures on the compact of recrystallized CBZ, not even after 8 h of DIDR measurement (Figure 4.5). Therefore, the first inflection point could not be correlated with the transformation of CBZ to its dihydrate. No second inflection point was determined.

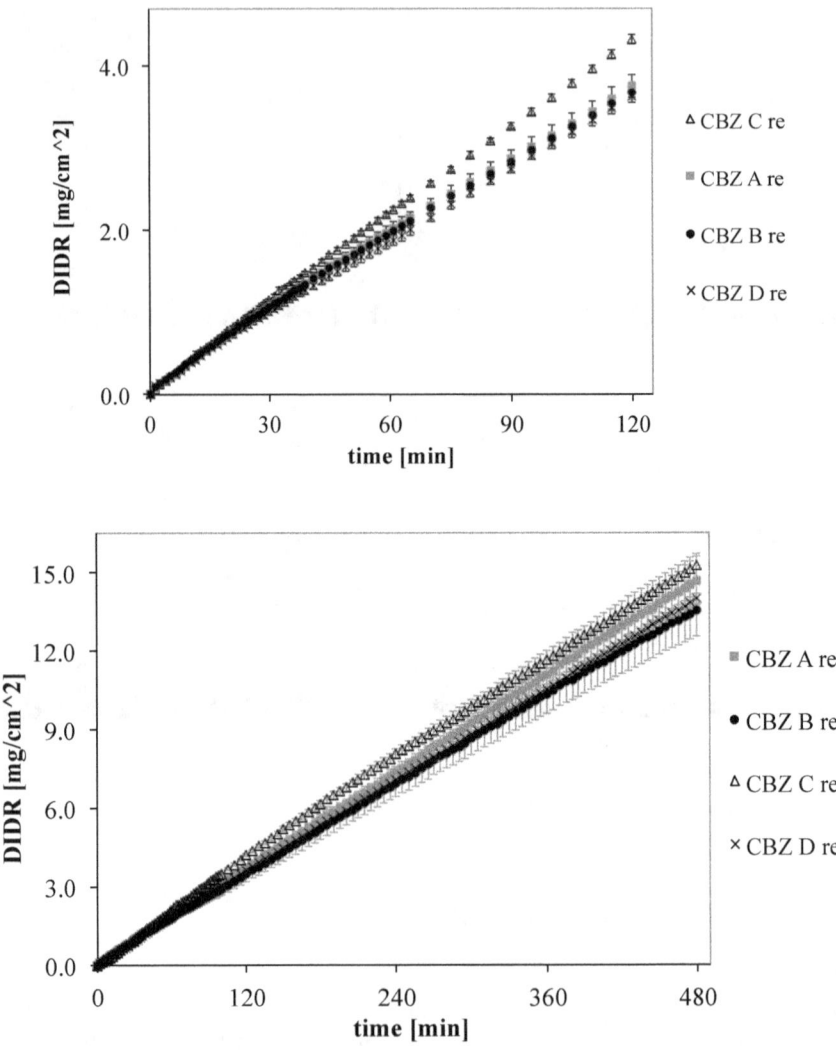

Figure 4.4: DIDR profiles of recrystallized CBZ A, B, C, and D; 2 h (left) and 8 h (right) DIDR test.

CBZ samples were recrystallized in presence of 1% PVP. Presence of PVP in recrystallization was reported to partially change the dissolution pattern of acetaminophen crystals (Wen et al., 2005). This change was associated with the adsorption of the polymer PVP on crystal surface by van der Waals interaction. PVP may have absorbed to the CBZ samples in the same manner. In addition, PVP is reported to inhibit transformation of CBZ to its dihydrate if present in the dissolution media at

Table 4.1: Inflection points in DIDR profiles of recrystallized CBZ samples.

Sample n = 3	First inflection point [min]
CBZ A re	31 ± 10
CBZ B re	28 ± 4
CBZ C re	24 ± 6
CBZ D re	21 ± 3

Figure 4.5: SEM picture of compact surface of recrystallized CBZ after 8 h DIDR test.

1% (Tian et al., 2006a). Although the amount of PVP absorbed on the recrystallized CBZ is very small (not detected by DSC or XRPD) it may be enough to explain the inhibited transformation found in the present results.

Properties of CBZ Compacts

Compacts of pure CBZ samples were difficult to obtain. There was a narrow compaction force range for each CBZ sample. Binding was insufficient below this range and above lamination and capping occurred. Best results were obtained for compaction forces 6.3–10.5 MPa for untreated and at much higher compaction force of 21.0 MPa for recrystallized CBZ samples. Compacts of recrystallized CBZ samples did not show any improvement in compaction as expected by the results reported by Nokhodchi et al. (2005). Nonetheless, the variability in tensile strength observed in compacts of recrystallized CBZ was reduced compared to tensile strength among compacts of untreated CBZ samples. Detailed results are shown in the Appendix 6.1.7.

Compared to anhydrous CBZ samples the CBZ dihydrate was of much easier to compact. At compaction forces of 6.3 MPa tensile strength was 1.921± 0.125 MPa.

Recrystallized CBZ in Binary Mixtures

Figure 4.6 shows the initial dissolution profiles of recrystallized CBZ in binary mixtures. The difference among recrystallized CBZ samples war reduced with the exception of the dissolution profiles of recrystallized CBZ D deviating significantly from the other dissolution profiles ($p < 0.05$). After 2 h dissolution, compacts showed swollen surfaces with wide creeks. This features were the strongest in MCC formulation with recrystallized CBZ D. It has to be considered that recrystallized CBZ samples were only slightly ground by mortar and pistil and particle size distribution was very wide. We therefore can not exclude inhomogeneous mixture in MCC formulations.

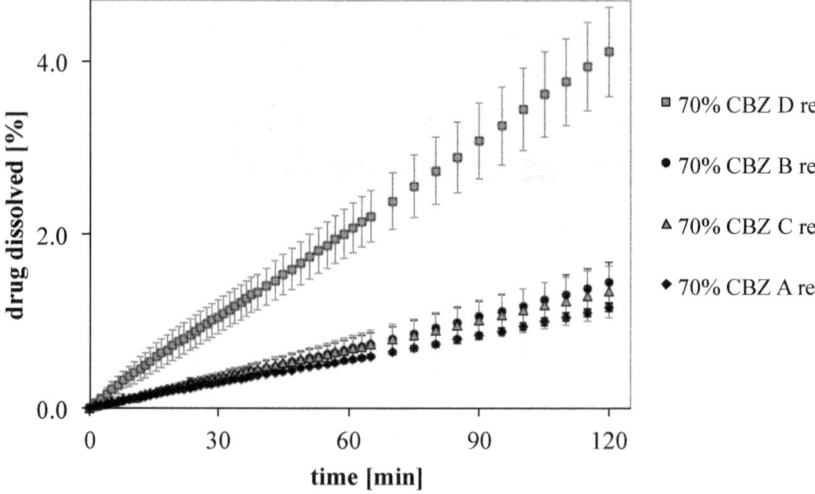

Figure 4.6: Initial drug release of recrystallized CBZ in binary mixtures with MCC.

Conclusions

Recrystallized CBZ samples presented a strongly reduced dissolution variability and the transformation to CBZ dihydrate was inhibited. Compacts of recrystallized CBZ samples also presented less variability in tensile strength compared to the untreated samples. The recrystallization did not induce any change in polymorphic form. Particle morphology of recrystallized CBZ samples was of irregular size and shape and may thus have resulted in inhomogeneous mixtures with MCC and to the deviation of one sample in the dissolution test. Therefore, more detailed studies are necessary to control the critical grinding step. The recrystallization in presence of 1% PVP can be suggested as a valid approach to control the dissolution variability in commercial CBZ samples. Although the results are promising, the careful selection of the CBZ raw material is suggested at the actual stage of knowledge.

Bibliography

Bolourtchian, N., Nokhodchi, A., Dinarvand, R., 2001. The effect of solvent and crystallization conditions on habit modification of carbamazepine. Daru 9, 12–22.

Flicker, F., Betz, G., 2012. Effect of crospovidone and hydroxypropyl cellulose on carbamazepine in high-dose tablet formulation. Drug Dev. Ind. Pharm. 4, 58–70.

Flicker, F., Eberle, V. A., Betz, G., 2011. Variability in commercial carbamazepine samples – impact on drug release. Int. J. Pharm. 410, 99–106.

Grzesiak, A. L., Lang, M., Kim, K., Matzger, A. J., 2003. Comparison of the four anhydrous polymorphs of carbamazepine and the crystal structure of form 1. J. Pharm. Sci. 92 (11), 2260–2271.

Krahn, F. U., Mielck, J. B., 1989. Effect of type and extent of crystalline order on chemical and physical stability of carbamazepine. Int. J. Pharm. 53, 25–34.

Lefebvre, C., Guyot-Hermann, A. M., Draguet-Brughmans, M., Bouché, R., Guyot, J. C., 1986. Polymorphic transitions of carbamazepine during grinding and compression. Drug Dev. Ind. Pharm. 12, 1913–1927.

Li, Y., Han, J., Zhang, G. G. Z., Grant, D. J. W., Suryanarayanan, R., 2000. In situ dehydration of carbamazepine dihydrate: A novel technique to prepare amorphous anhydrous carbamazepine. Pharm. Dev. Technol 5, 257–266.

Lindenberg, M., Kopp, S., Dressman, J. B., 2004. Classification of orally administered drugs on the world health organization model list of essential medicines according to the biopharmaceutics classification system. Eur. J. Pharm. Biopharm. 58, 265–278.

Mahalaxmi, R., Ravikumar, Pandey, S., Shirwaikar, A., Shirwaikar, A., 2009. Effect of recrystallization on size, shape, polymorph and dissolution of carbamazepine. Int. J. PharmTech Res. 1 (3), 725–732.

McMahon, L. E., Timmins, P., Williams, A. C., York, P., 1996. Characterization of dihydrates prepared from carbamazepine polymorphs. J. Pharm. Sci. 85 (10), 1064–1069.

Mosharraf, M., Sebhatu, T., Nyström, C., 1999. The effects of disordered structure on the solubility and dissolution rates of some hydrophilic, sparingly soluble drugs. Int. J. Pharm. 177, 29–51.

Murphy, D., Rodríguez-Cintrón, F., Langevin, B., Kelly, R. C., Rodríguez-Hornedo, N., 2002. Solution-mediated phase transformation of anhydrous to dihydrate carbamazepine and the effect of lattice disorder. Int. J. Pharm. 246, 121–134.

Nokhodchi, A., Bolourtchian, N., Dinarvand, R., 2005. Dissolution and mechanical behaviors of recrystallized carbamazepine from alcohol solution in the presence of additives. J. Cryst. Growth 274, 573–584.

Rodrìguez-Sponga, B., Priceb, C. P., Jayasankara, A., Matzger, A. J., Rodrìguez-Hornedo, N., 2004. General principles of pharmaceutical solid polymorphism: a supramolecular perspective. Adv. Drug Deliv. Rev. 56, 241–274.

Rustichelli, C., Gamberini, G., Ferioli, V., Gamberini, M., Ficarra, R., Tommasini, S., 2000. Solid-state study of polymorphic drugs: carbamazepine. J. Pharm. Biomed. Anal. 23, 41–54.

Tian, F., Sandler, N., Gordon, K. C., McGoverin, C. M., Reay, A., Strachan, C. J., Saville, D. J., Rades, T., 2006a. Visualizing the conversion of carbamazepine in aqueous suspension with and without the presence of excipients: A single crystal study using SEM and Raman microscopy. Eur. J. Pharm. Biopharm. 64, 326–335.

Tian, F., Zeitler, J., Strachan, C., Saville, D., Gordon, K., Rades, T., 2006b. Characterizing the conversion kinetics of carbamazepine polymorphs to the dihydrate in aqueous suspension using Raman spectroscopy. J. Pharm. Biomed. Anal. 40, 271–280.

Šehić, S., Betz, G., Hadžidedić, Š., El-Arini, S. K., Leuenberger, H., 2010. Investigation of intrinsic dissolution behavior of different carbamazepine samples. Int. J. Pharm. 386, 77–90.

Wen, H., Morris, K. R., , Park, K., 2005. Study on the interactions between polyvinylpyrrolidone (PVP) and acetaminophen crystals: Partial dissolution pattern change. J. Pharm. Sci. 94, 2166–2174.

Chapter 5

Dissolution Project with the Optimized CBZ Tablet Formulation

Abstract

Although drug release of the tablet formulation developed earlier was within the USP specification using various CBZ samples, the dissolution curves differed with the CBZ sample used. The aim of this project was to study drug release of the tablet formulation in water without any additives. Drug release of CBZ tablet formulation with CBZ A, B, C, and D was analyzed by flow-through cells. In water, the tablet formulation showed a robust drug release regarding the tested CBZ samples. Moreover, flow-through cells in open dissolution mode can be suggested to study effects of various dissolution media on poorly soluble drugs like CBZ.

Introduction

CBZ tablets have a history of irregular dissolution behavior and bioinequivalence (Mittapalli et al., 2008; Davidson, 1995; Meyer et al., 1992). In water and at high relative humidity CBZ transforms to the less soluble CBZ dihydrate (Rodrìguez-Sponga et al., 2004). This transformation can be critical to the dissolution behavior of the final tablet formulation. Šehić et al. (2010) reported different transformation within commercial CBZ samples and this difference also shows in tablet formulation using Ludipress® as tablet filler.

To remove the effect of individual transformation a CBZ tablet formulation was developed selecting the two excipients crospovidone (CrosPVP) and hydroxypopyl cellulose (HPC), both reported to inhibit transformation (Flicker and Betz, 2012). The dissolution conformed to the requirements of the USP monograph for CBZ tablets of immediate release. Testing different CBZ raw materials in the tablet formulation the dissolution curves were still within the USP specifications, yet showing significant difference.

The dissolution medium described in USP monograph of CBZ tablets requires 1% (w/v) of the surfactant sodium laurylsulfate (SLS) in water. SLS strongly enhances the solubility of CBZ and sink conditions can be maintained in a closed system with 900 ml dissolution medium and 280 mg CBZ. However, SLS is reported to enhance also the transformation to CBZ dihydrate (Rodrìguez-Hornedo and Murphy, 2004; Luhtala, 1992). Therefore, analysis of CBZ release in presence of SLS may not show the ability of the excipients CrosPVP and HPC to reduce the dissolution variability.

In this project, the robustness of CBZ tablet formulation towards various commercial CBZ samples was tested in water without any further additives. Therefore, flow-through cells were selected to test the CBZ tablet formulation in open dissolution mode providing sink conditions. Drug dissolution was compared to the dissolution method described in the USP monograph for CBZ tablet of immediate release.

Materials and Methods

CBZ Tablet Formulation

The CBZ tablet formulation developed within the PhD studies (Section 3.2) was used. The tablet composition was 70% CBZ (API), 18% MCC (filler), 6% HPC-SL (binder), 5% CrosPVP (superdisintegrant), and 1% magnesium stearate (lubricant). Tablet formulation was prepared using CBZ raw materials of four different suppliers, namely CBZ A, B, C, and D. Tablets were of 400 mg, flat-faced, 10-mm in diameter, $12.1 \pm 0.2\%$ in porosity, and prepared by direct compaction using a simulated speed of 0.5 m/s and 6–18 kN compaction force.

Flow-Through Cells (USP 4)

Dissolution test was performed by flow-through cells, the USP Apparatus 4 (Sotax*CE7-smart*, SOTAX AG, Switzerland). The 22.6 mm cells were used in laminar mode (packed column with 1 mm glass beads) and tablets were placed in a tablet holder. Dissolution was performed in the open system mode, at 16 ml/min flow rate, 37 ± 0.5 °C, and in dissolution medium of water and 1% SLS, respectively. Samples were analyzed by UV-VIS Spectrophotometer (Lambda 25,

PerkinElmer) at 285 nm and 287 nm respective to the wavelength of maximum absorption (λ_{max}) of CBZ in water and 1% SLS.

Results and Discussion

Figures 5.1 and 5.2 show the dissolution curves of the CBZ tablet formulation in water and in 1% SLS. Flow-through cells in open dissolution mode present a small volume of fresh dissolution medium to the tablet at continuous flow. With the fast disintegrating CBZ tablet formulation this small volume dissolved a high amount of CBZ in the initial dissolution phase. The measured CBZ concentration decreased over the dissolution time (Subfigures 5.1a and 5.2a).

Mean dissolution curves of CBZ tablets with CBZ A, B, C, and D were compared by one-way ANOVA. While dissolution curves of CBZ tablets analyzed in 1% SLS were significantly different ($p < 0.001$), CBZ tablets revealed no significant difference among tablets with CBZ A, B, C, and D when analyzed in water ($p > 0.1$).

Figure 5.3 shows the dissolution curves of CBZ tablets with CBZ A, B, C, and D obtained by the paddle method using 1% SLS as dissolution medium (same results as shown in Section 3.2). These dissolution curves were compared to the ones obtained by the flow through cells using same dissolution medium (Figure 5.2 (b)). Both showed amount of drug released that was within the USP specification, where at 15 min amount of CBZ released is 45–75% and at 60 min more than 75%. Amount of drug dissolved at t_{15} and t_{60} of the different methods are shown in Table 5.1. Also the analysis by one-way ANOVA showed comparable results, mean dissolution curves of CBZ tablets with CBZ A, B, C, and D differed significantly in dissolution data obtained by both dissolution methods ($p < 0.01$). However, the precision of dissolution curves obtained by the flow through cells was lower (RSD $\leq 8.6\%$) compared to the one obtained by paddle method (RSD $\leq 2.9\%$). At CBZ concentrations above 640 mg/L in 1% SLS the correlation with the UV absorption was not linear anymore. This may explain the especially high standard deviation observed for the tablet formulation of CBZ B (Figure 5.2 (b)). Nonetheless, the dissolution variability was comparable.

Conclusions

Excluding the effect of SLS on the dissolution, the tablet formulation was robust regarding CBZ raw material of supplier A, B, C, and D. Flow through cells presented the advantage of keeping sink condition throughout the full dissolution testing by using the open dissolution mode. The flow through cells provide a valuable tool to study the effect of different dissolution media on poorly soluble drug like CBZ.

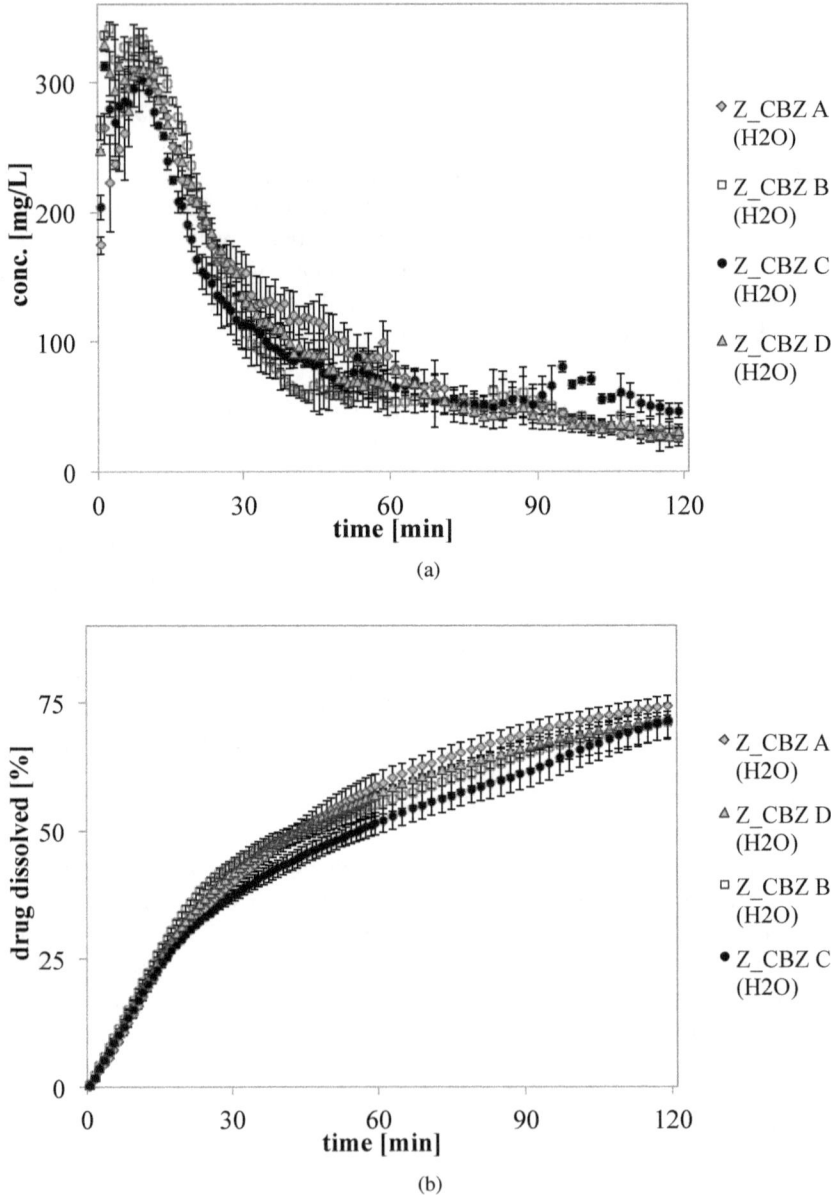

Figure 5.1: Dissolution of CBZ tablet formulation with CBZ A, B, C, and D in water; drug concentration (a) and cumulative amount of drug dissolved [%] (b) over time.

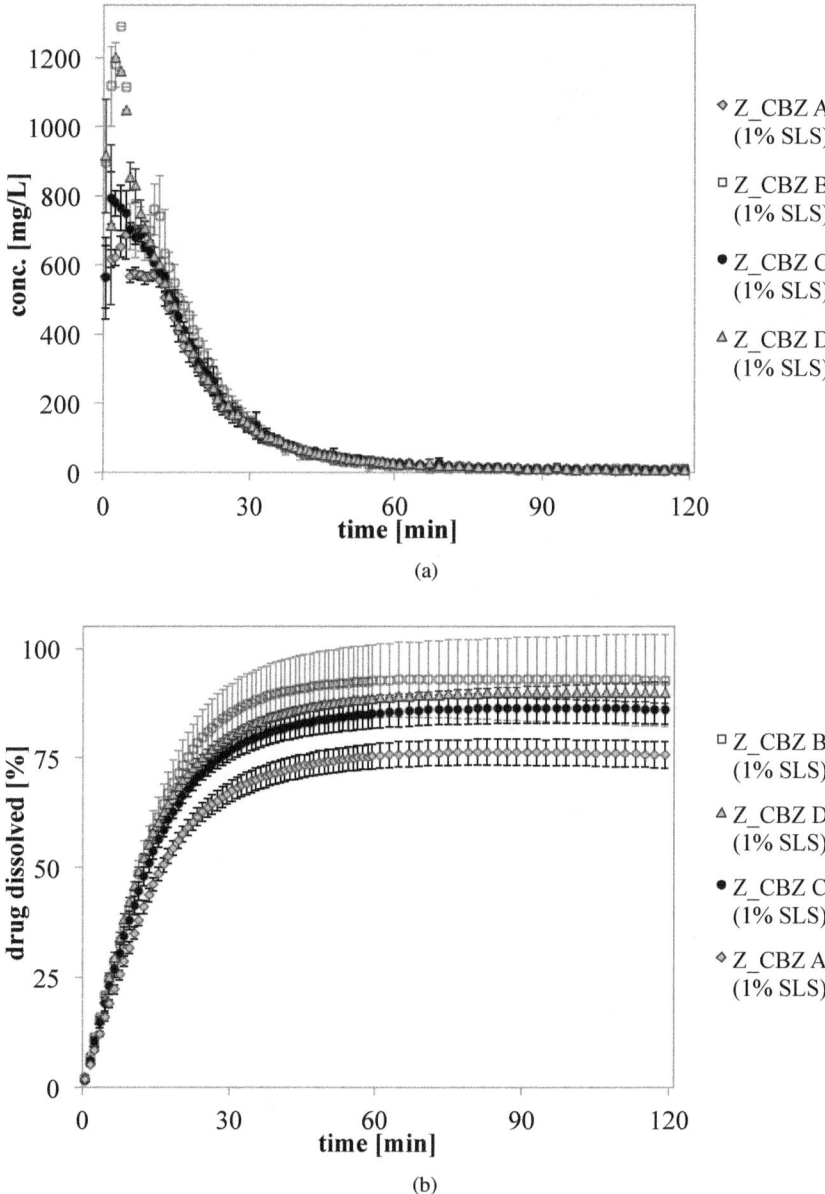

Figure 5.2: Dissolution of CBZ tablet formulation with CBZ A, B, C, and D in 1% SLS; drug concentration (a) and cumulative amount of drug dissolved [%] (b) over time.

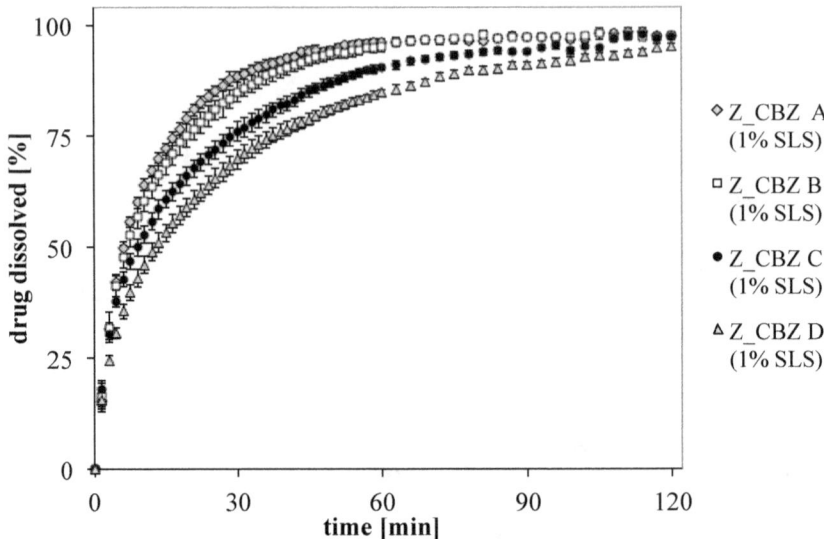

Figure 5.3: Dissolution curves of CBZ tablets analyzed by the paddle method in 1% SLS (results of Section 3.2).

Table 5.1: Amount of CBZ dissolved at t_{15} and at t_{60}; CBZ tablets with CBZ A, B, C, and D dissolved in water and in 1% SLS using flow-through cells (FTC) and paddle method.

$n \geq 3$	H_2O (FTC)	1% SLS (FTC)	1% SLS (Paddle)
CBZ		dd_{15} [%]	
A	23.3 ± 0.8	47.4 ± 1.6	72.1 ± 1.4
B	26.6 ± 0.5	60.2 ± 4.4	68.8 ± 2.6
C	23.5 ± 0.2	55.0 ± 1.9	60.8 ± 2.0
D	25.0 ± 0.2	59.2 ± 0.7	53.2 ± 2.9
CBZ		dd_{60} [%]	
A	58.8 ± 2.5	75.4 ± 2.8	96.1 ± 0.7
B	55.2 ± 2.3	92.6 ± 8.4	95.1 ± 1.1
C	51.5 ± 1.4	85.1 ± 3.3	90.5 ± 0.6
D	57.1 ± 0.8	88.5 ± 0.5	84.8 ± 2.9

Bibliography

Davidson, A., 1995. A multinational survey of the quatity of carbamazepine tablets. Drug Dev. Ind. Pharm. 21, 2167–2186.

Flicker, F., Betz, G., 2012. Effect of crospovidone and hydroxypropyl cellulose on carbamazepine in high-dose tablet formulation. Drug Dev. Ind. Pharm. 4, 58–70.

Luhtala, S., 1992. Effect of sodium lauryl sulphate and polysorbate 80 on crystal growth and aqueous solubility of carbamazepine. Acta Pharm. Nordica 4, 1100–1801.

Meyer, M. C., Straughn, A. B., Jarvi, E. J., Wood, G. C., Pelsor, F. R., Shah, V. P., 1992. The bioinequivalence of carbamazepine tablets with a history of clinical failures. Pharm. Res. 9, 1612–1616.

Mittapalli, P. K., Suresh, B., Hussaini, S. S. Q., Rao, Y. M., Apte, S., 2008. Comparative in vitro study of six carbamazepine products. AAPS PharmSciTech 9 (2), 357.

Rodrìguez-Hornedo, N., Murphy, D., 2004. Surfactant-facilitated crystallization of dihydrate carbamazepine during dissolution of anhydrous polymorph. J. Pharm. Sci. 93, 449–460.

Rodrìguez-Sponga, B., Priceb, C. P., Jayasankara, A., Matzger, A. J., Rodrìguez-Hornedo, N., 2004. General principles of pharmaceutical solid polymorphism: a supramolecular perspective. Adv. Drug Deliv. Rev. 56, 241–274.

Šehić, S., Betz, G., Hadžidedić, Š., El-Arini, S. K., Leuenberger, H., 2010. Investigation of intrinsic dissolution behavior of different carbamazepine samples. Int. J. Pharm. 386, 77–90.

Chapter 6

Conclusions

During this study, the variability in CBZ raw materials of four different suppliers was characterized and its impact on drug release was studied. The CBZ samples showed differences in polymorphic purity, particle morphology, and intrinsic dissolution behavior. Most interestingly, two inflection points were found in the intrinsic dissolution profiles that characterized the transformation behavior of each anhydrous CBZ samples to the CBZ dihydrate as a time range. Furthermore, the applied unidirectional dissolution method allowed investigating the influence of commonly used tablet fillers already at preformulation level. Presence of MCC reduced the drug release variability found for the CBZ samples and can be suggested as a suitable tablet filler for CBZ.

Recrystallizing CBZ samples in presence of 1% PVP resulted in strongly reduced dissolution variability and the transformation to CBZ dihydrate was inhibited. Nonetheless, considering the high amount of solvent, the low yield of 53–70%, and the critical grinding step following the recrystallization, the careful selection of the CBZ raw material is suggested at the actual stage of knowledge.

At the formulation level, high-dose CBZ tablets of immediate release were successfully prepared with the novel combination of crospovidone (CrosPVP) as superdisintegrant, hydroxypropyl cellulose (HPC-SL) as dry binder, and MCC as the tablet filler. The optimal formulation was with 5% CrosPVP and 6% HPC, where tablet formulation of the four CBZ raw materials conformed to the USP requirements for CBZ immediate release tablets. Moreover, drug release in water was robust towards the variability in the CBZ samples showing no significant difference among the dissolution curves.

In conclusion, the strategy suggested to control the variability in CBZ samples includes the unidirectional dissolution method as a straightforward monitoring tool in preformulation studies. To allow a certain variability in CBZ raw materials, it is further suggested to design a robust formulation incorporating the excipients CrosPVP, HPC, and MCC.

Building on the presented studies, it would be illuminative to investigate the robustness of the CBZ tablet formulation in in-vivo conditions. Furthermore, it would be of great interest to validate the unidirectional dissolution method on another hydrate-forming drug like theophylline.

Appendix

The Appendix provides additional information on the original publications and the recrystallization project. Section 6.1 presents additional information on Publication 1 and on the recrystallization project, whereas Section 6.2 presents further information on Publication 2 and on the dissolution project.

6.1 Additional Information on Publication 1 and on the Recrystallization Project

6.1.1 Further Analyses on CBZ Samples

Light Microscopy

Morphology of CBZ samples was analyzed by an inverse light microscope (Hund Wetzlar Wilovert®, Germany). Images were taken by a camera (Canon Digital IXUS 80 IS, Japan) and through the SPL Ph 20/0.25 objective (Figure 6.1).

The images reveal different morphology of the CBZ samples with very crystalline particles of prismatic shape in CBZ A, particles with rough and uneven edges and a lot of fines in CBZ B and D, and more round particles with smooth surface in CBZ C.

Same morphological structure can be reported using the more sophisticated technique, the SEM (Section 3.1). Nevertheless, light microscopy presents a fast and easy to use tool with very little sample preparation.

Thermogravimetry

CBZ A, B, and D were analyzed by thermogravimetry (Pyris TGA 6, Perkin Elmer, USA). Samples of 8–15 mg were scanned from 40–200 °C at 10 °C/min with nitrogen gas purge of 100 ml/min. Loss of mass was registered on a microbalance. Thermogravimetric analysis was conducted in triplicate. Mass loss was $0.12 \pm 0.08\%$, $0.09 \pm 0.02\%$, and $0.09 \pm 0.01\%$ for CBZ A, B, and D, respectively. The

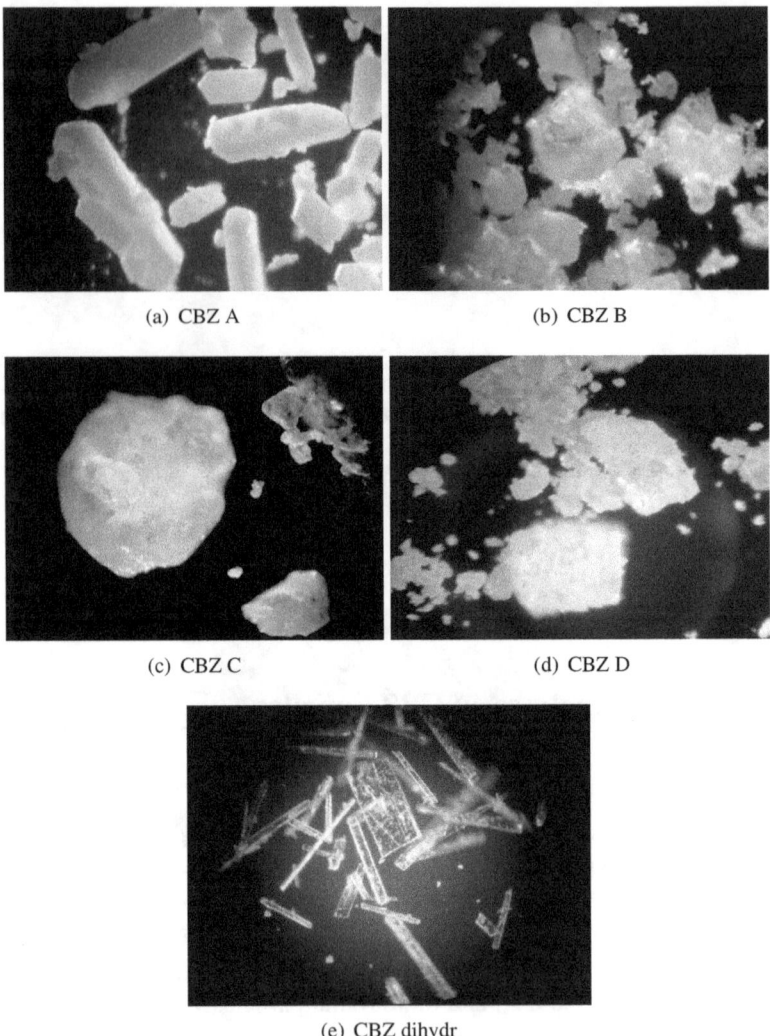

Figure 6.1: Images of CBZ A, B, C, and D and CBZ dihydrate by light microscopy.

values present no substantial weight loss and it can be concluded that the CBZ samples did not contain any solvent. In contrast, CBZ dihydrate showed a mass loss of 9.87 ± 0.20% (Figure 6.2). The theoretical stoichiometric water content of CBZ dihydrate is 13.2% w/w. The obtained mass loss was of lower value and indicates that CBZ dihydrate was not fully hydrated. Hence, the IDR values of fully hydrated CBZ are expected to present even lower values than detected (Section 3.1).

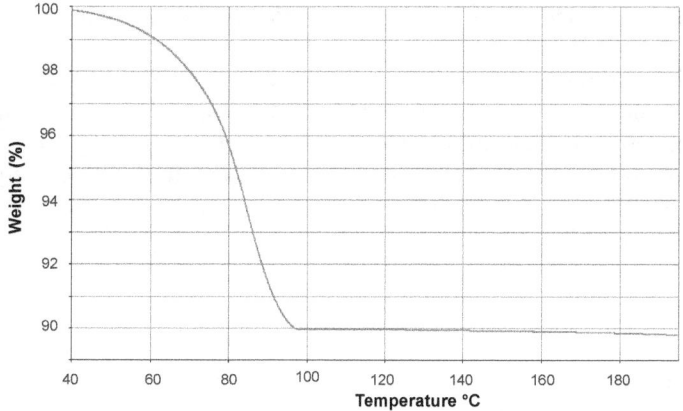

Figure 6.2: Weight loss under heating of CBZ dihydrate.

Flowability of CBZ

Flowability of CBZ samples was determined according to USP 31 (2008) using a tap volumeter (type STAV 2003, J. Engelsmann AG, Germany). A powder sample of 100 g was poured into a graduated cylinder. The initial volume (ml) was recorded for the bulk density in g/cm^3 and the volume after tapping 1250 times was recorded for the tap density. Flowability test was performed in triplicate.

Table 6.1: Flowability of CBZ samples: bulk and tapped density, Hausner ratio, and Compressibility Index.

CBZ n = 3	bulk density [g/cm^3]	tapped density [g/cm^3]	Hausner ratio	Compressibility Index [%]
A	0.61 ± 0.00	0.68 ± 0.00	1.12 ± 0.01	10.4 ± 1.0
B	0.68 ± 0.00	0.78 ± 0.00	1.16 ± 0.00	13.5 ± 0.0
N	0.76 ± 0.01	0.85 ± 0.00	1.11 ± 0.01	9.7 ± 0.8
D	0.76 ± 0.01	0.84 ± 0.01	1.11 ± 0.00	10.1 ± 0.3

The results indicate good flow for all CBZ samples, Hausner ratio was below 1.25 and Compressibility Index (CI) was below 16%. CBZ N even showed a CI of less than 10% indicating excellent flow properties. Therefore, flowability is not expected to limit manufacturing of the CBZ samples.

6.1.2 Development of the Unidirectional Dissolution Method

The unidirectional dissolution method was developed based on the manual disk intrinsic dissolution rate (DIDR) method by Šehić et al. (2010). The unidirectional dissolution method allows the drug to dissolve from one compact surface only. If the compact is of pure drug, the unidirectional dissolution method is equal to an automatic DIDR method. In the following section the automatic DIDR is compared to the manual DIDR method.

Experimental

Sample Preparation. Flat-faced compacts of 400 mg and 0.95 cm^2 surface area were prepared from CBZ A, B, and D using a material tester (Zwick 1478, Germany). Compact porosity was controlled to 12% and hardness was 20–41 N. For compacts of CBZ A, porosity was 16%, as higher compaction forces resulted in lamination. Compaction force was set between 6–10 kN, while compaction, decompaction, and ejection speed were set at 10 mm/min, 50 mm/min, and 10 mm/min, respectively. After elastic recovery the compacts were placed into the sample holder (disk) and embedded with melted paraffin wax so only one surface was available to the dissolution media.

Manual DIDR Method. Manual DIDR was carried out in 400 ml water at 37 °C. Disk rotation was 100 rpm and the dissolution medium was additionally stirred by a magnetic bar. Aliquots of 5 ml were withdrawn every minute during the first 20 min, every 5 min up to 40 min, every 10 min up to 60 min, and every 20 min up to a total of 120 min dissolution time. Dissolution medium was replaced after every sampling. Six samples were analyzed simultaneously and the disks were lifted out of the medium during the sampling and replacement of dissolution medium. Drug content of the samples was analyzed by UV spectrophotometer (DU 530, Beckman, USA) at 285 nm.

Automatic DIDR Method. Automatic DIDR method was performed at 100 rpm in 500 ml water at 37 °C. A commercially available dissolution machine (Sotax*AT7smart*) was modified at its rotating basket unit to a rotating disk. The Sotax*AT7smart* is a closed system with automatic drug content measurement at predetermined time intervals. Initial measurement interval was set at 1.3 min during 39 min, then at 2 min up to 65 min, and at 5 min up to a total of 120 min dissolution time. The interval of 1.3 min was found to be the smallest possible measuring interval, where sufficient pumping (1.5 times turnover of the medium in the sampling loop) and no system conflict occurred. Drug content was measured by UV-VIS spectrophotometer (Lambda 25, PerkinElmer, USA) at 285 nm.

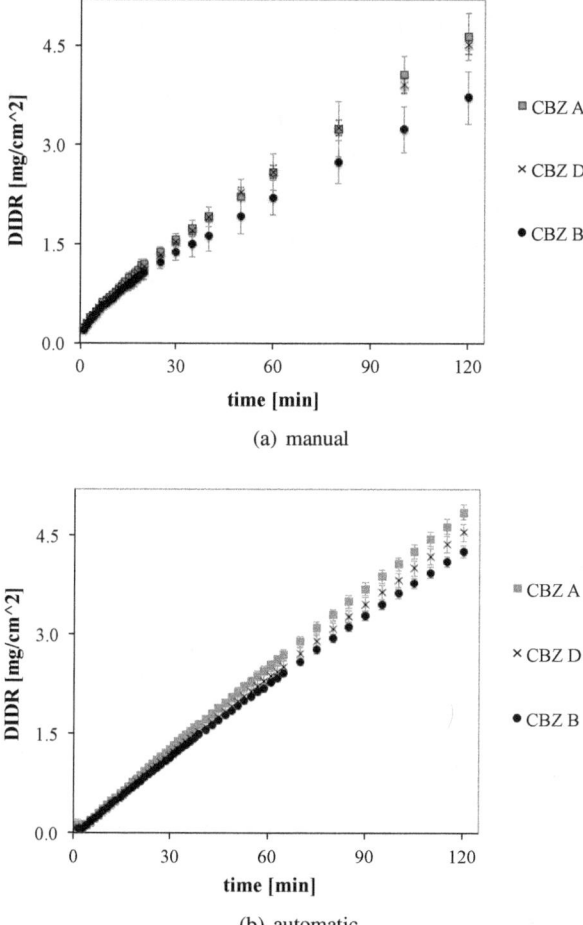

Figure 6.3: DIDR profiles of the manual (a) and automatic (b) DIDR method.

Results and Discussion

Figure 6.3 shows the DIDR profiles of the manual (a) and the automatic (b) DIDR method, respectively. The DIDR profiles of the manual method show a clear inflection around 20 min, and DIDR profiles of the automatic method show comparatively low standard deviation.

IDR values were determined by the slope in the DIDR profile prior to the inflection point, the slope referring to the release of anhydrous CBZ before transforming to the dihydrate. IDR values and their standard deviation (SD) are shown in Table

Table 6.2: Intrinsic dissolution rate (IDR) values obtained by the manual and the automatic DIDR method.

Sample n ≥ 5	[$\mu g/cm^2/min$]	
	IDR manual	IDR automatic
CBZ A	53.0 ± 10	43.9 ± 3.1
CBZ B	49.5 ± 5.3	38.6 ± 0.8
CBZ D	45.0 ± 2.7	39.9 ± 0.7

6.2. Both, IDR and SD values were higher in the manual method compared to the automatic one.

Inflection point. The inflection point was determined by the intercept of two linear regressions. First regression line included the initial DIDR profile approximated to the best fit and second regression line was set through all data points of the later stage in the DIDR profile that were not included in first regression line. Best fit in linear regression was R^2 0.9973 ± 0.0026 and R^2 0.9997 ± 0.0003 for the initial DIDR profiles of the manual and automatic DIDR method, respectively.

Table 6.3: Inflection point (IP) by the manual and the automatic DIDR method.

Sample n ≥ 5	IP manual [min]	IP automatic [min]
CBZ A	16 ± 6	33 ± 6
CBZ B	14 ± 5	28 ± 10
CBZ D	29 ± 4	41 ± 7

DIDR profiles obtained from the automatic DIDR method showed the inflection points in the same order as of the manual DIDR method: CBZ B > A > D. However, inflection points in DIDR profiles obtained by the automatic method were at a later time. The manual DIDR method is disruptive and the DIDR profiles do not present real time contact with the dissolution media. This can explain the later time of inflection points as well as the higher IDR values in DIDR profiles obtained by the automatic method.

Compact surfaces during the DIDR measurement. When the disks are lifted out of the dissolution medium for the sample measurement a stagnant film of water stays on the disk surface. Over-saturation can occur and thereby enhance the transformation to CBZ dihydrate and the needles can grow without the convective force in the dissolution medium. Figure 6.4 shows the compact surfaces during the manual DIDR measurement, whereas Figures 6.5 are the compact surfaces during the automatic mode. In the automatic DIDR method the dihydrate crystals show at a later time and as much smaller needles (note the different sampling times and different resolutions).

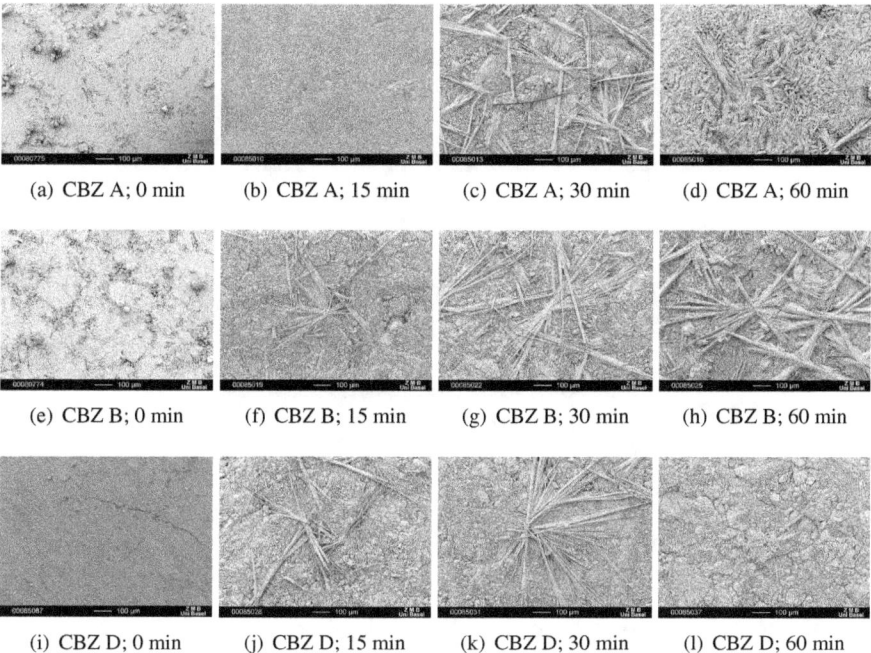

(a) CBZ A; 0 min   (b) CBZ A; 15 min   (c) CBZ A; 30 min   (d) CBZ A; 60 min

(e) CBZ B; 0 min   (f) CBZ B; 15 min   (g) CBZ B; 30 min   (h) CBZ B; 60 min

(i) CBZ D; 0 min   (j) CBZ D; 15 min   (k) CBZ D; 30 min   (l) CBZ D; 60 min

Figure 6.4: SEM pictures of compact surface from CBZ A, B, and D at 0, 15, 30, and 60 min manual DIDR test; magnification of $\times 100$.

Conclusions

The results confirm that the manual DIDR method allows the distinction between CBZ samples of different supplier by an inflection point in the DIDR profiles reflecting transformation to the CBZ dihydrate. However, transformation was found to be enhanced by the disruptive test settings, which results in an inflection point visible without further calculations. The drawbacks of the manual methods are the precision, the manual sampling, and the time limitation thereby. The automatic DIDR method presents a system with continuous dissolution, higher precision, and the ability to analyze the samples over a longer time range.

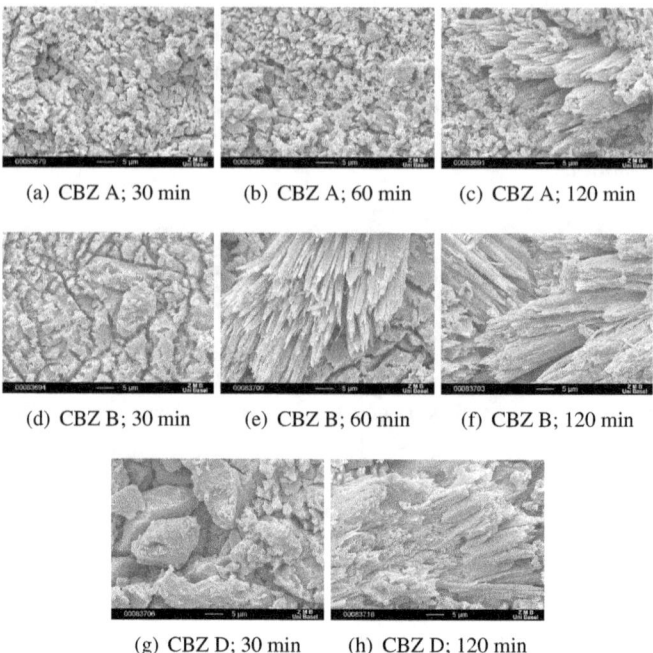

Figure 6.5: SEM pictures of compact surface from CBZ A, B, and D at 30, 60, and 120 min automatic DIDR test; magnification of ×2000.

6.1.3 Precision and Effect of Particle Size

The effect of particle size on the precision of DIDR profiles was analyzed in the manual DIDR method. Sieve fractions (125 – 355 µm) of CBZ A, B, and D were compared with the non-sieved samples (n = 3). Precision of DIDR profiles is described by the coefficient of variation (CV), i.e., the relative standard deviation,

$$CV\,[\%] = \frac{SD}{\bar{x}} \cdot 100 \tag{6.1}$$

where SD is the standard deviation and $\bar{x}$ is the arithmetic mean. The average relative standard deviation over the full DIDR profile was calculated. Surprisingly DIDR profiles of sieved fractions showed with $13.4 \pm 2.8\%$ the higher relative standard deviation (RSD) compared to the non-sieved samples with RSD of $7.2 \pm 1.4\%$. Moreover, IDR values of sieved CBZ were higher than of non-sieved CBZ samples (73.5 ± 8.1 and 69.2 ± 5.1 mg/cm^2/min). The physical strain by the sieving process may have induced some crystal defects and thereby led to the inhomogeneous and faster dissolution of sieved CBZ samples. Therefore, non-sieved CBZ samples were used in all further experiments.

DIDR profiles in the unidirectional dissolution method showed an average RSD of less than 4%, with the exception of CBZ D, where DIDR profiles showed average RSD values up to 10%. In general, RSD was higher in the initial DIDR profiles (up to around 20 min) and was reduced to RSD values less than 2% after 2 h DIDR test.

6.1.4 Repeatability

Repeatability of IDR values and inflection points of CBZ samples was tested by a second data set (n = 5). The unidirectional dissolution method was conducted by a different person (master student V. A. Eberle). The results are compared to the previous data set (Section 3.1).

Figure 6.6 shows the DIDR profile of the 2 h and 11 h-run. IDR values of the initial phase and the two inflection points are shown in Table 6.4 and 6.5, respectively.

Table 6.4: Intrinsic dissolution rate of CBZ samples in the initial DIDR profiles prior to the first inflection point.

Sample n ≥ 5	IDR [$\mu g/cm^2/min$]
CBZ A	37.2 ± 1.0
CBZ B	33.6 ± 1.8
CBZ C	36.9 ± 1.7
CBZ D	32.3 ± 1.7

Table 6.5: First and second inflection point in DIDR profiles of CBZ samples.

Sample n ≥ 3	First inflection point (min)	Second inflection point (min)
CBZ A	35 ± 6	546 ± 26
CBZ B	23 ± 6	321 ± 52
CBZ C	19 ± 2	334 ± 196
CBZ D	26 ± 6	555 ± 26

Initial phase in DIDR profiles showed a significant difference in IDR for CBZ samples (ANOVA $p < 0.001$). The IDR values were in the same range as of the first data set. However, IDR values were not distinct among each other and could not be ranked in the same order as reported for the previous data set.

CBZ samples differed significantly in their first and second inflection point (ANOVA $p < 0.005$). First inflection points could be ranked as CBZ C≤B<D<A ($p < 0.05$),

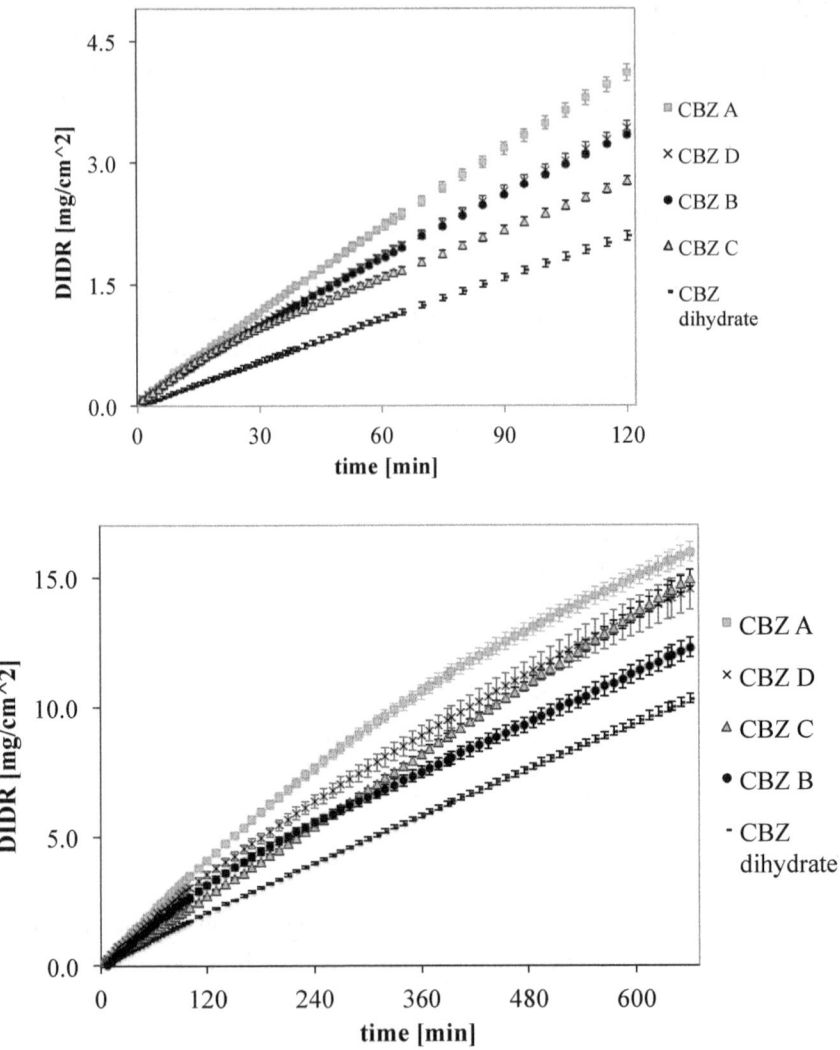

Figure 6.6: DIDR profiles of CBZ A, B, C, and D with the 2 h (top) and 11 h (bottom) dissolution test. Average values (n = 5) are presented.

and second inflection points as CBZ C$\leq$B<A$\leq$D ($p < 0.0005$). The ranking was the same as for the previous data set. However, the ranking did not show the same siginificance.

Conclusions. IDR and inflection points in DIDR profiles were successfully repeated. The range of the first and second inflection points, their precision, and

(a) 0 h   (b) 8 h

Figure 6.7: SEM image of CBZ dihydrate, compact surface before and after 8 h DIDR test.

their ranking were confirmed by a different data set measured by a different person. Also the range of IDR values was confirmed, however, with no significant ranking. These findings lead to the conclusion, that the inflection points present a adequate parameter to characterize variability among commercial CBZ sample of different suppliers.

6.1.5 Inflection Point in DIDR Profiles of CBZ Dihydrate?

DIDR profiles also showed an inflection point for CBZ dihydrate (around 35 min). The reason for this phenomenon is most likely the onset of crystal growth and rearrangement of the dihydrate crystals to bigger crystals aligned in block-like forms. For visualization, see compact surface of CBZ dihydrate after 8 h DIDR test in Figure 6.7.

The inflection point in DIDR profiles of CBZ dihydrate does not represent the transformation process as found for the anhydrous CBZ samples. The inflection point for CBZ dihydrate is therefore not discussed in Publication 1.

6.1.6 Transformation of CBZ to Dihydrate – Contradictory Results in Literature

As soon as anhydrous CBZ transforms to its dihydrate, the dissolution rate is expected to decrease because of the lower solubility of CBZ dihydrate (Kaneniwa et al., 1987; Kobayashi et al., 2000). Interestingly, Tian et al. (2007) reported contrary results. In presence of excipients inhibiting the transformation, CBZ samples presented a decreased dissolution rate compared to the dissolution rate in water. The hypothesis was that the formed dihydrate crystals presented a bigger surface area and thereby enhanced the drug dissolution. Within the PhD study, the decrease in dissolution rate correlated with the appearance of the needle-like dihydrate crystals on the compact surfaces.

Transformation of CBZ anhydrous to dihydrate can be induced by physical processes like grinding leading to disrupted crystal lattice and finally to a chasnged dissolution behavior (Murphy et al., 2002; Lefebvre et al., 1986). However, the significance of the effect on dissolution is reported confusingly. Tenho et al. (2007) compared IDR values between ground and unground CBZ and reported no significant difference. The IDR values were calculated excluding the initial 30 min of dissolution and linear regression was placed through the following linear phase. Moreover, the considered dissolution range showed two clearly distinct curves that seem to be parallel to each other and the dissolution profile of the ground CBZ samples was at a higher concentration compared to the unground CBZ samples. Therefore, it has to be concluded, that the ground CBZ samples showed a faster drug release in the initial phase and thereby clearly showing an effect of grinding on drug release.

6.1.7 Compact Hardness of Untreated and Recrystallized CBZ Samples

Table 6.6 shows the compaction force necessary to compact untreated and recrystallized CBZ samples to a fixed porosity and the tensile strength of the obtained compacts.

Table 6.6: Compaction force (CF) and tensile strength (TS) of untreated and recrystallized CBZ samples. Compacts were of fixed porosity.

$n \geq 5$	untreated CBZ		recrystallized CBZ	
	CF [MPa]	TS [MPa]	CF [MPa]	TS [MPa]
CBZ A	10.5	0.355 ± 0.036	21.0	0.630 ± 0.098
CBZ B	8.4	0.505 ± 0.127	21.0	0.556 ± 0.262
CBZ C	6.3	0.294 ± 0.017	21.0	0.436 ± 0.252
CBZ D	8.4	0.651 ± 0.095	21.0	0.498 ± 0.180
$\bar{x} \pm SD$	8.4	0.451 ± 0.160	21.0	0.530 ± 0.083

6.1.8 Effects of Mannitol and MCC on CBZ

Effect of Mannitol. Binary mixtures of CBZ with mannitol showed an interesting phenomenon on the compact surface during the DIDR test. A new type of crystal shape was detected on the surface arranged like the spines of a hedgehog (Figure 6.8).

(a) x 100 (b) x 500

Figure 6.8: SEM pictures of compact surface from 70% CBZ B in binary mixtures with mannitol after 120 min DIDR test.

The question arises of which molecules these crystals are formed. Are they of mannitol, CBZ, or of a cocrystal? Mannitol itself exhibits polymorphism and it dissolves much faster than CBZ. Hence, the crystals could be of another less soluble polymorphic form of mannitol. A further and more likely possibility is that mannitol acts as impurity in the solution-mediated transformation of CBZ and thereby influence the crystal habit of CBZ dihydrate. It would be very interesting to analyze these crystals as to their chemical composition and polymorphic form.

Binary mixtures of CBZ with high amount of mannitol (30% drug load) showed an increased initial dissolution; the amount of drug dissolved was of up to 10 times higher compared to the binary mixtures of higher drug load. The high solubility of mannitol may provide a reason, it dissolves much faster than CBZ and leaves a porous structure of the less soluble CBZ with increased surface area. The porous structure could be shown on the compact surface during the DIDR test by SEM imaging (Figure 6.9).

Figure 6.9: SEM pictures of compact surface from 30% CBZ B in binary mixtures with mannitol after 120 min DIDR test.

Figures 6.10 and 6.11 show the diffractograms of CBZ samples in binary mixtures with mannitol. No interaction of mannitol with CBZ could be detected. The diffractogram of mannitol revealed, that the polymorphic form of the preprocessed Parteck® M300 is mannitol form I according to Burger et al. (2000).

Effect of MCC. Binary mixtures of CBZ with MCC showed a clear reduction in dissolution variability. This is surprising considering the compact surface after 120 min DIDR test (Figure 6.12), where compact surfaces visibly differ with the CBZ sample. A possible explanation for the reduced variability could be that the fiber-like structure of MCC stabilized CBZ release.

Diffrctograms of CBZ samples in binary mixtures with MCC are shown in Figures 6.13 and 6.14. No interaction could be detected.

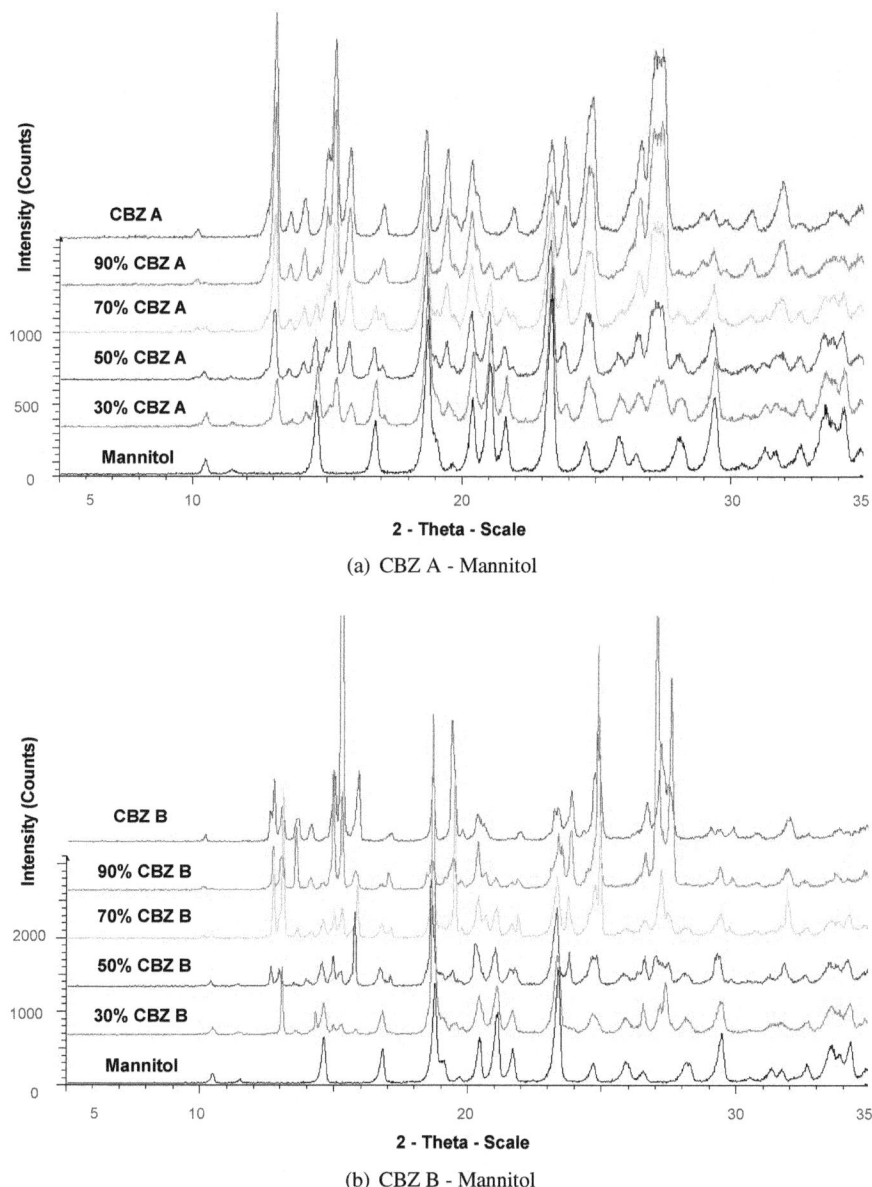

Figure 6.10: XRPD of CBZ A and B in binary mixtures with MCC (0–90% drug load) and pure mannitol.

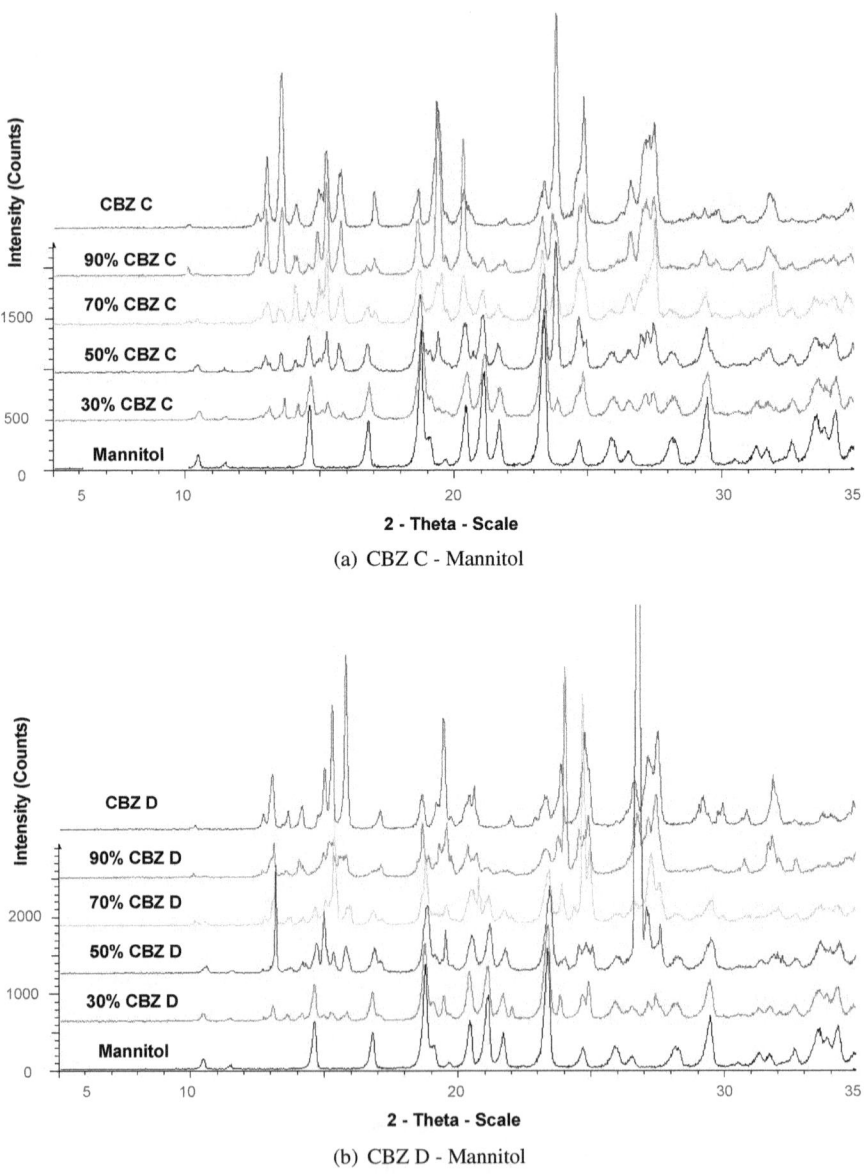

Figure 6.11: XRPD of CBZ C and D in binary mixtures with MCC (0–90% drug load) and pure mannitol.

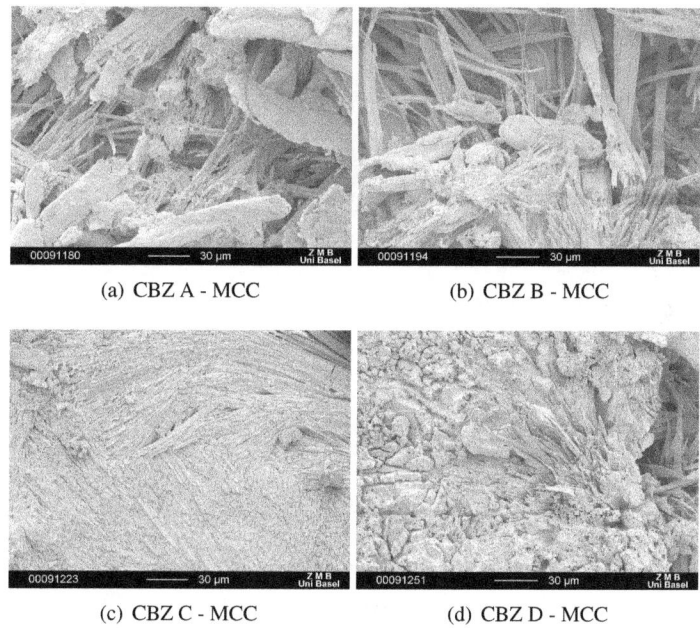

Figure 6.12: SEM pictures of compact surface from 70% CBZ in binary mixtures with MCC after 120 min DIDR test.

6.1.9 UV Calibration

Various UV calibration curves were necessary to analyze the CBZ concentration dependent on the required sensitivity range and on the system used (Table 6.7).

Table 6.7: Calibration curves of CBZ in different dissolution media of type $y = ax + b$.

Medium	range[a]	λ_{max}[b]	cell	a	b	R^2
H_2O	0.1 – 20	285	10 mm[c]	0.0508	-0.0094	0.99988
H_2O	1 – 15	285	10 mm[d]	0.05108	0.00970	0.99986
H_2O	15 – 150	285	1 mm[d]	0.00515	-0.00003	0.99917

[a] in [mg/L]
[b] in [nm]
[c] UV-VIS spectrophotometer, Beckman, USA
[d] UV-VIS spectrophotometer, Lambda 25, PerkinElmer, USA

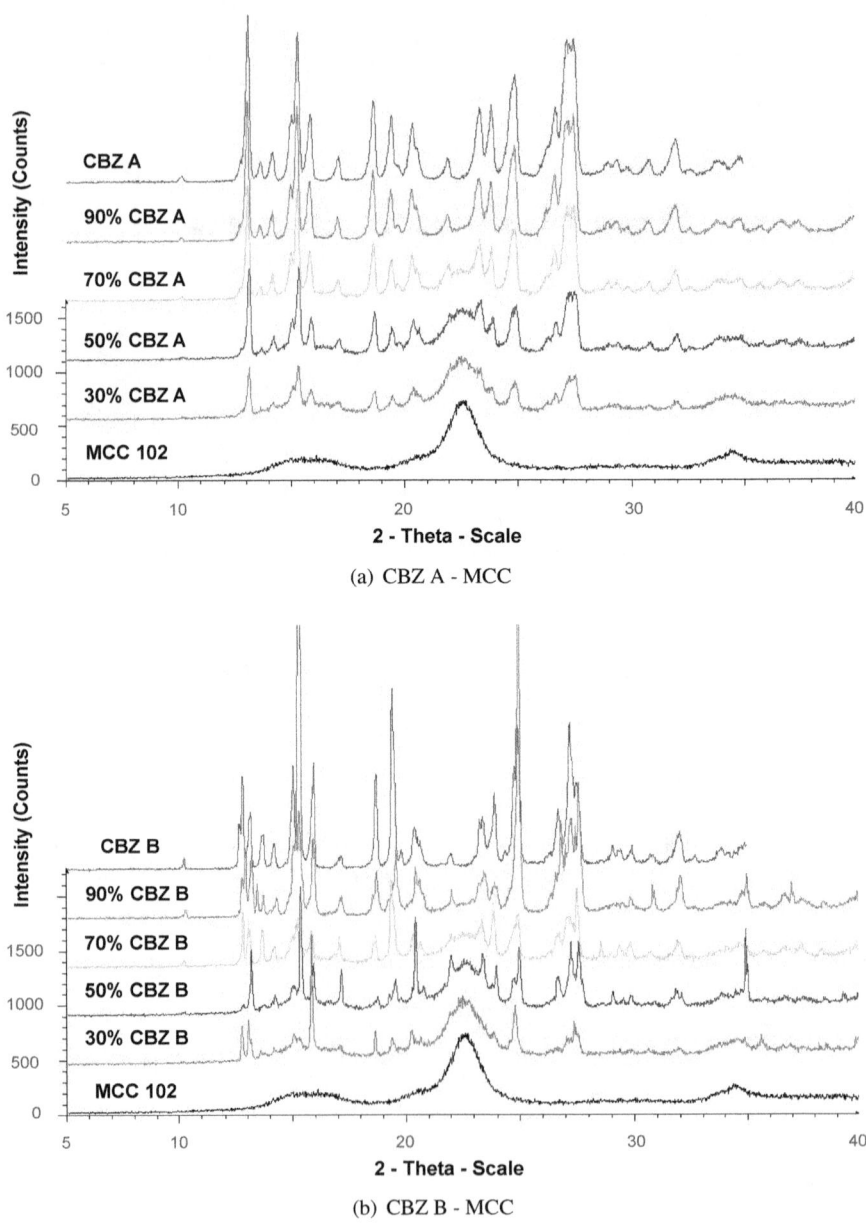

Figure 6.13: XRPD of CBZ A and B in binary mixtures with MCC (0–90% drug load) and pure MCC.

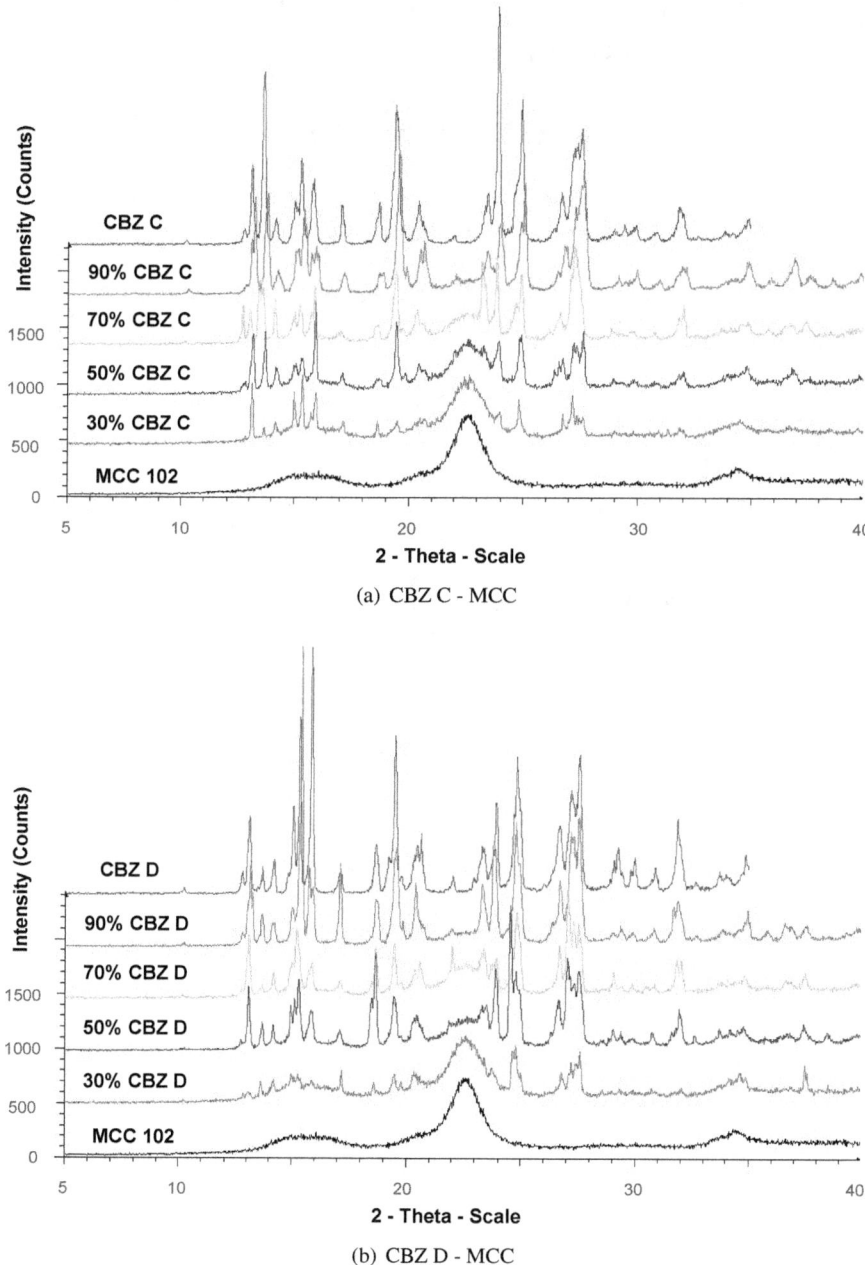

Figure 6.14: XRPD of CBZ C and D in binary mixtures with MCC (0–90% drug load) and pure MCC.

6.2 Additional Information on Publication 2 and on the Dissolution Project

6.2.1 SEM Images of the Excipients in the Tablet Formulation

SEM images of the excipients used for the high-dose CBZ tablets are shown in Figure 6.15. The particle morphology differs with the function of the excipient. CrosPVP presents very porous particles indicating the ability to absorb water fast and act as a superdisintegrant. The filler-binder MCC shows aggregated fibrous structure, same can be said for the binder HPC-SL, although particle shape is less defined.

(a) MCC (b) CrosPVP (c) HPC-SL

Figure 6.15: SEM images of filler-binder MCC, superdisintegrant CrosPVP, and binder HPC-SL

6.2.2 UV Calibration

Depending on the dissolution medium and the required sensitivity range different UV calibration curves were acquired using the UV-VIS spectrophotometer (Lambda 25, Perkin-Elmer, USA). The parameters of the various calibration curves are shown in Table 6.8.

Table 6.8: Calibration curves of CBZ in different dissolution media of type $y = ax + b$.

Medium	range[a]	λ_{max}[b]	cell	a	b	R^2
1% SLS	11 – 300	287	1 mm	0.00491	-0.00042	0.99999
1% SLS	10 – 640	287	1 mm	0.00456	0.04420	0.99911
H$_2$0	25 – 250	285	1 mm	0.00551	0.00429	0.99996

[a] in [mg/L]
[b] in [nm]

The dissolution method by the flow through cells required UV calibration of higher CBZ concentrations to analyzed the initial dissolution profiles. However, in the dissolution medium containing 1% SLS CBZ concentrations of $\geq$ 640 mg/L did not show linear correlation with the UV absorption and were therefore not included.

Bibliography

Burger, A., Henck, J.-O., Hetz, S., Rolling, J. M., Weissnicht, A. A., Stöttner, H., 2000. Energy/temperature diagram and compression behavior of the polymorphs of d-mannitol. J. Pharm. Sci. 89, 457–468.

Kaneniwa, N., Ichikawa, J.-I., Yamaguchi, T., Hayashi, K., Watari, N., Otsuka, M., 1987. Dissolution behavior of carbamazepine polymorphs. Yakugaku Zasshi 107, 808–813.

Kobayashi, Y., Ito, S., Itai, S., Yamamoto, K., 2000. Physicochemical properties and bioavailability of carbamazepine polymorphs and dihydrate. Int. J. Pharm. 193, 137–146.

Lefebvre, C., Guyot-Hermann, A. M., Draguet-Brughmans, M., Bouché, R., Guyot, J. C., 1986. Polymorphic transitions of carbamazepine during grinding and compression. Drug Dev. Ind. Pharm. 12, 1913–1927.

Murphy, D., Rodríguez-Cintrón, F., Langevin, B., Kelly, R. C., Rodríguez-Hornedo, N., 2002. Solution-mediated phase transformation of anhydrous to dihydrate carbamazepine and the effect of lattice disorder. Int. J. Pharm. 246, 121–134.

Tenho, M., Heinänen, P., Tanninen, V. P., Lehto, V.-P., 2007. Does the preferred orientation of crystallites in tablets affect the intrinsic dissolution? J. Pharm. Biomed. Anal. 43, 1315–1323.

Tian, F., Sandler, N., Aaltonen, J., Lang, C., Saville, D. J., Gordon, K. C., Strachan, C. J., Rantanen, J., Rades, T., 2007. Influence of polymorphic form, morphology, and excipient interactions on the dissolution of carbamazepine compacts. J. Pharm. Sci. 96 (3), 584–594.

USP 31, 2008. USP 31; United States Pharmacopoeia / National Formulary. United States Pharmacopeial Convention.

Šehić, S., Betz, G., Hadžidedić, Š., El-Arini, S. K., Leuenberger, H., 2010. Investigation of intrinsic dissolution behavior of different carbamazepine samples. Int. J. Pharm. 386, 77–90.